Second Edition

Structural and Stress Analysis

Theories, Tutorials and Examples

Second Edition

Structural and Stress Analysis

Theories, Tutorials and Examples

Jianqiao Ye
Lancaster University
United Kingdom

CRC Press
Taylor & Francis Group
Boca Raton London New York

CRC Press is an imprint of the
Taylor & Francis Group, an **informa** business

CRC Press
Taylor & Francis Group
6000 Broken Sound Parkway NW, Suite 300
Boca Raton, FL 33487-2742

© 2016 by Taylor & Francis Group, LLC
CRC Press is an imprint of Taylor & Francis Group, an Informa business

No claim to original U.S. Government works

Printed on acid-free paper
Version Date: 20151019

International Standard Book Number-13: 978-1-4822-2033-9 (Paperback)

This book contains information obtained from authentic and highly regarded sources. Reasonable efforts have been made to publish reliable data and information, but the author and publisher cannot assume responsibility for the validity of all materials or the consequences of their use. The authors and publishers have attempted to trace the copyright holders of all material reproduced in this publication and apologize to copyright holders if permission to publish in this form has not been obtained. If any copyright material has not been acknowledged please write and let us know so we may rectify in any future reprint.

Except as permitted under U.S. Copyright Law, no part of this book may be reprinted, reproduced, transmitted, or utilized in any form by any electronic, mechanical, or other means, now known or hereafter invented, including photocopying, microfilming, and recording, or in any information storage or retrieval system, without written permission from the publishers.

For permission to photocopy or use material electronically from this work, please access www.copyright.com (http://www.copyright.com/) or contact the Copyright Clearance Center, Inc. (CCC), 222 Rosewood Drive, Danvers, MA 01923, 978-750-8400. CCC is a not-for-profit organization that provides licenses and registration for a variety of users. For organizations that have been granted a photocopy license by the CCC, a separate system of payment has been arranged.

Trademark Notice: Product or corporate names may be trademarks or registered trademarks, and are used only for identification and explanation without intent to infringe.

Visit the Taylor & Francis Web site at
http://www.taylorandfrancis.com

and the CRC Press Web site at
http://www.crcpress.com

To my wife Qin and my daughter Helen,
with love and gratitude

Contents

Preface to the first edition

This book is not intended to be an additional textbook of structural and stress analysis for students who have already been offered many excellent textbooks that are available in the market. Instead of going through rigorous coverage of the mathematics and theories, this book summarises major concepts and important points that should be fully understood before students can claim to have successfully completed the subject. One of the main features of this book is that it aims at helping students to understand the subject through asking and answering conceptual questions, in addition to solving problems based on applying the derived formulas.

It has been found that by the end of a Structural and Stress Analysis course most of our students are able to follow the instructions given by their lecturers and can solve problems if they can identify suitable formulas. However, they may not necessarily fully understand what they are trying to solve and what is really meant by the solution they have obtained. For example, they may have found the correct value of a stress, but may not understand what is meant by 'stress'. They may be able to find the direction of a principal stress if they know the formula, but may not be able to give a rough prediction of the direction without carrying out a calculation. To address these issues, understanding all the important concepts of structures and stresses is essential. Unfortunately, this has not been appropriately highlighted in the mainstream textbooks since the ultimate task of these textbooks is to establish the fundamental theories of the subject and to show the students how to derive and use the formulas.

Each chapter of this book begins with a summary of key issues and relevant formulas. This is followed by a key points review that identifies important concepts essential for students' understanding of the chapter. Next, numerical examples are used to illustrate these concepts and the application of the formulas. A short discussion of the problem is always provided before following the solution procedure to ensure that students know not only how but also why a formula should be used in such a way. Unlike most of the textbooks available in the market, this book asks students to answer only questions that require minimum or no numerical calculations. Questions requiring extensive numerical calculations are not duplicated here since they can be easily found from other textbooks. The conceptual questions ask students to review important concepts and test their understanding of the concepts. These questions can also be used by lecturers to organise group discussions in the class. At the end of each chapter, there is a mini test including both conceptual and numerical questions.

This book is written to be used with a textbook of your choice, as a useful companion. It is particularly useful when students are preparing for their examinations. Asking and answering these conceptual questions and reviewing the key points summarised in this book is a structured approach to assess whether or not the subject has been understood and to identify the area where further revision is needed. The book is also a useful reference for those who are taking an advanced structural and stress analysis course. It provides a quick recovery of the theories and important concepts that have been learnt in the past, without the need to pick up those from a much detailed and, indeed, thicker textbook.

The author is indebted to the external reviewers appointed by the publishers and is appreciative of their constructive comments and criticism. This book would not have been completed without the support received from the School of Civil Engineering, the University of Leeds, where I have been employed as an academic member of staff.

The interest of Spon Press in the publication of this book is greatly appreciated. I would like to thank Tony Moore, Matthew Gibbons, Monika Falteiskova and Katy Low for their encouragement and editorial assistance.

Preface to the second edition

The publication of this second edition of *Structural and Stress Analysis* was encouraged by the enthusiastic response to the first edition that I received from students and lecturers.

I am fortunate to have the opportunity to teach either civil or mechanical engineering degree students at various universities. This has prompted me to revise the first edition of this book to broaden the content so that some of the advanced topics can be included for a wider spectrum of readers. The new additions include plastic deformation of beams, bending of thin plates and a brief introduction to impact and vibration.

The greater part of the second edition, however, comprises materials of the first edition, and for this I am most grateful to all those who were involved in its production. The continuous interest of Spon Press in the publication of this book is greatly appreciated. I would especially like to thank Tony Moore for his encouragement and support, without which this edition would not have been completed. I am also grateful for the help I have received from Kathryn Everett in preparing the final version of the manuscript.

Chapter 1

Introduction

Any material or structure may possibly fail when it is loaded. The successful design of a structure requires detailed structural and stress analysis in order to assess whether or not it can safely support the required loads. Figure 1.1 shows how a structure behaves under applied loads.

To prevent structural failure, a typical design must consider the following three major aspects:

1. *Strength* – The structure must be strong enough to carry the applied loads.
2. *Stiffness* – The structure must be stiff enough such that only allowable deformation occurs.
3. *Stability* – The structure must not collapse through buckling subjected to compressive loads.

The subject of structural and stress analysis provides analytical, numerical and experimental methods for determining the strength, stiffness and stability of load-carrying structural members.

1.1 FORCE AND MOMENT

A *force* is a measure of its tendency to cause a body to move or translate in the direction of the force. A complete description of a force includes its *magnitude* and *direction*. The magnitude of a force acting on a structure is usually measured by Newton (N), or kilo-newton (kN). In stress analysis, a force can be categorised as either external or internal. External forces include, for example, applied surface loads, force of gravity and support reactions. Internal forces are the resisting forces generated within loaded structural elements. Typical examples of applied external forces include the following:

a. Point load, where a force is applied through a point of a structure (Figure 1.2a)
b. Distributed load, where a force is applied over an area of a structure (Figure 1.2b)

The *moment* of a force is a measure of its tendency to cause a body to rotate about a specific point or axis. In order to develop a moment about, for example, a specific axis, a force must act such that the body would begin to twist or bend about the axis. The magnitude of the moment of a force acting about a point or axis is directly proportional to the distance

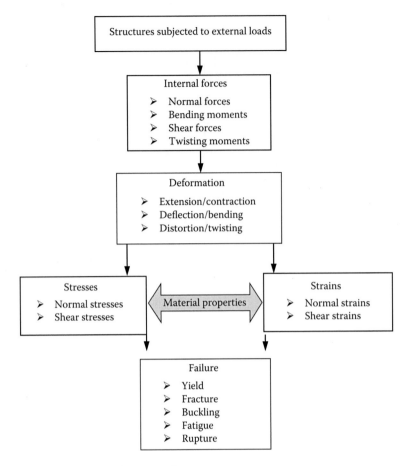

Figure 1.1 Flow chart of structural and stress analysis.

of the force from the point or axis. It is defined as the product of the force and the *lever arm*. The lever arm is the perpendicular distance between the line of action of the force and the point about which the force causes rotation. A moment is usually measured in Newton-meters (N m), or kilo-newton-meters (kN m). Figure 1.3 shows how a moment about the beam–column connection is caused by the applied point load F.

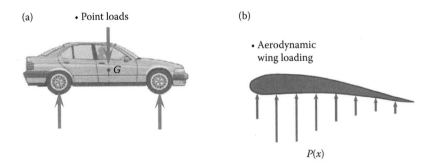

Figure 1.2 Examples of external loadings.

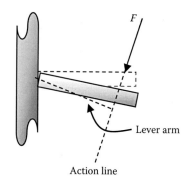

Figure 1.3 Moment of a force.

1.2 TYPES OF FORCE AND DEFORMATION

1.2.1 Force

On a cross section of a material subjected to external loads, there exist four different types of internal forces (Figure 1.4):

1. Normal force, F, which is perpendicular to the cross section
2. Shear force, V, which is parallel to the cross section
3. Bending moment, M, which bends the material
4. Twisting moment (torque), T, which twists the material about its central axis

1.2.2 Deformation

Table 1.1 shows the most common types of forces and their associated deformations. In a practical design, the deformation of a member can be a combination of the basic deformations shown in Table 1.1.

1.3 EQUILIBRIUM SYSTEM

In static structural and stress analysis, a system in equilibrium implies that:

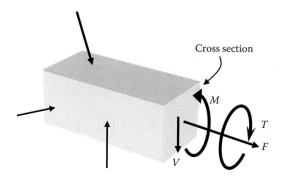

Figure 1.4 Internal forces on a cross section.

Table 1.1 Basic types of deformation

Force	Deformation	Description
Normal force Axial force Thrust		The member is stretched by the axial force and is in **tension**. The deformation is characterised by axial **elongation**.
Normal force Axial force Thrust		The member is compressed by the axial force and is in **compression**. The deformation is characterised by axial **shortening**.
Shear force		The member is sheared. The deformation is characterised by **distorting** a rectangle into a parallelogram.
Torque Twist moment		The member is twisted and is in **torsion**. The deformation is characterised by **angle of twist**.
Bending moment		The member is bent and is in **bending**. The deformation is characterised by a **bent shape**.

- The resultant of all applied forces, including support reactions, must be zero.
- The resultant of all applied moments, including bending and twisting moments, must be zero.

The two equilibrium conditions are commonly used to determine support reactions and internal forces on the cross sections of structural members.

1.3.1 Free body diagram of an object or system

One of the most important tools for calculating support reactions is the use of *free body diagram* (FDB). The principle of drawing an FBD of a system is to identify all the forces, including any externally applied forces, loads, self-weight, support reactions and forces transferred from any neighbouring bodies, acting on the system and to investigate how the equilibrium is maintained under the action of all the forces. The following steps summarise how an FBD can be drawn for the system shown in Figure 1.5:

1. Make sure that you know exactly for which part of the system you want to draw an FBD, for example, for the car, the bridge or the car and bridge together.
2. Isolate the part (body) from the system.
3. Identify all the externally applied forces, including loads, self-weight, etc.
4. Identify where the body contacts other parts of the system.

Figure 1.5

5. Define the types of reaction forces at those contacts.
6. Draw the externally applied forces and the reactions.

Table 1.2 demonstrates how the FBDs of the car and the bridge are drawn by following the six steps listed below.

1.3.2 Method of section

One of the most basic analyses is the investigation of the internal resistance of a structural member, that is, the development of internal forces within the member to resist the effect of the externally applied forces. The *method of section* is normally used for this purpose. Table 1.3 shows how the method of section works.

Table 1.2 Free body diagram

Step	FBD of	
1	Car	Bridge
2		
3	Self-weight	Self-weight
4	Contacts with the bridge at the wheels	1. Contacts with the car at the wheels 2. Contacts with the ground at the two ends
5	1. Vertical upwards reaction forces from the bridge acting at the contacts 2. Vertical downwards self-weight of the car	1. Vertical downwards reaction forces at the car-bridge contacts 2. Vertical upwards reaction forces at the bridge-ground contacts 3. Vertical downwards self-weight of the bridge
6		

Table 1.3 The method of section

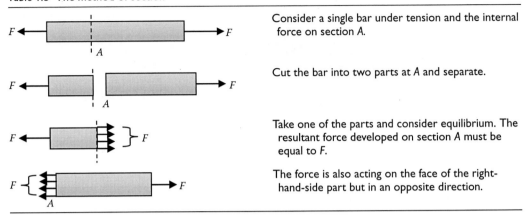

	Consider a single bar under tension and the internal force on section A.
	Cut the bar into two parts at A and separate.
	Take one of the parts and consider equilibrium. The resultant force developed on section A must be equal to F.
	The force is also acting on the face of the right-hand-side part but in an opposite direction.

In summary, if a member as a whole is in equilibrium, any part of it must also be in equilibrium. Thus, the externally applied forces acting on one side of an arbitrary section must be balanced by the internal forces developed on the section.

1.3.3 Method of joint

The analysis or design of a truss requires the calculation of the internal forces in each of its members. Taking the entire truss as a free body, the forces in the members are internal forces. In order to determine the internal forces in the members jointed at a particular joint, for example, joint A in Figure 1.6, the joint can be separated from the truss system by cutting all the members around it. On the sections of the cuts there exist axial forces that can be further determined by considering the equilibrium of the joint under the action of all the internal forces and the externally applied loads at the joint, that is, by resolving all the forces in the x and y directions, respectively, and letting the resultants be zero.

1.4 STRESSES

Stress can be defined as the intensity of internal force, representing internal force per unit area at a point on a cross section. Stresses are usually different from point to point. There are two types of stresses, namely *normal* and *shear stresses*.

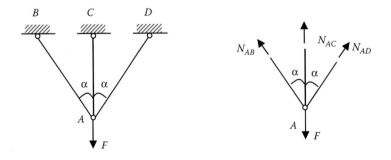

Figure 1.6 Equilibrium of truss.

1.4.1 Normal stress

Normal stress is a stress perpendicular to a cross section or cut. For example, for the simple structural element shown in Figure 1.7a, the normal stress on section $m-m$ can be calculated as

$$\text{Normal stress } (\sigma) = \frac{\text{force (on section } m-m)}{\text{area (of section } m-m)} \tag{1.1}$$

The basic unit of stress is N/m², which is also called Pascal.

In general, a stress varies from point to point (Figure 1.7b). A general stress can be calculated by

$$\text{Stress at point } P = \lim_{\Delta A \to 0} \frac{\Delta F}{\Delta A} \tag{1.2}$$

where ΔF is the force acting on the infinitesimal area, ΔA, surrounding P.

1.4.2 Shear stress

Shear stress is a stress parallel to a cross section or cut. For example, for the plates connected by a bolt, shown in Figure 1.8a, the forces are transmitted from one part of structure to the other by causing stresses in the plane parallel to the applied forces. These stresses are shear stresses. To compute the shear stresses, a cut is taken through the parallel plane and uniform distribution of the stresses over the cut is assumed. Thus

$$\tau = \frac{\text{force}}{\text{area under shearing}} = \frac{P}{A} \tag{1.3}$$

where A is the cross-sectional area of the bolt.

(a)

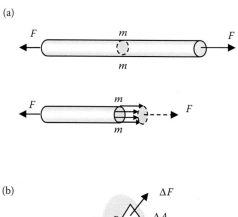

(b)

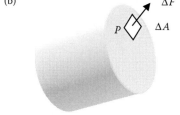

Figure 1.7 Normal stress on a cross section.

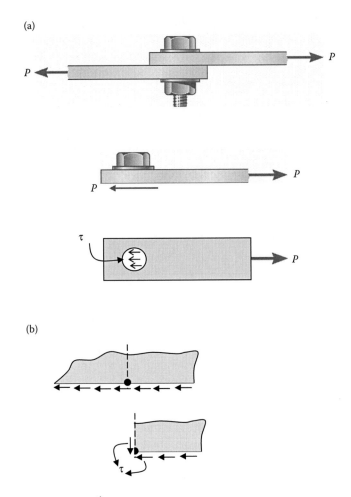

(a)

(b)

Figure 1.8 Shear stress on a cross section.

At a point in a material, shear stresses always appear in pair acting on two mutually perpendicular planes. They are equal in magnitude, but in an opposite sense, that is, either towards or away from the point (Figure 1.8b).

From the definition of normal and shear stresses, the following three characteristics must be specified in order to define a stress:

1. The magnitude of the stress
2. The direction of the stress
3. The plane (cross section) on which the stress is acting

1.5 STRAINS

Strain is a measure of relative deformation. Strains can be categorised as normal and shear strains.

Figure 1.9 Illustration of normal strain.

Normal strain is a measure of the change in length per unit length under stress (Figure 1.9). It is measured by the following formula:

$$\text{Normal strain}(\varepsilon) = \frac{\text{change in length}}{\text{original length}} = \frac{\Delta L}{L} \tag{1.4}$$

Shear strain is a measure of the distortion caused by shear stresses in the right angle between two fibres within a plane (Figure 1.10). It is measured by the tangent of the two small angles, α_i and α_2, that is,

$$\text{Shear strain } \gamma = \text{Tan}(\alpha_1) + \text{Tan}(\alpha_2) \tag{1.5}$$

Since the change of angles is very small, shear strain can be calculated approximately by

$$\gamma \approx \alpha_1 + \alpha_2 \tag{1.6}$$

Strains are dimensionless.

1.6 STRAIN–STRESS RELATION

The relationship between strain and stress depends on the properties of materials. For a linear elastic material, strain–stress relation is also termed as *Hooke's law*, which determines how much strain occurs under a given stress. For materials undergoing linear elastic

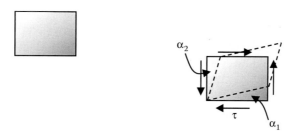

Figure 1.10 Illustration of shear strain.

deformation, stresses are proportional to strains. Thus, for the simple load cases shown in Figures 1.9 and 1.10, the strain–stress relations are

$$\sigma = E\varepsilon$$
$$\tau = G\gamma$$

$$(1.7)$$

where E is called *modulus of elasticity* or *Young's modulus*. G is termed as *shear modulus*. They are all material-dependent constants and have the unit of stress, for example, N/mm^2, since strains are dimensionless. For isotropic materials, for example, most metals, E and G, have the following relationship:

$$G = \frac{E}{2(1 + v)}$$

$$(1.8)$$

In Equation 1.8, v is called *Poisson's ratio*, which is also an important material constant. Figure 1.11 shows how a Poisson's ratio is defined by comparing axial elongation and lateral contraction of a prismatic bar in tension. Poisson's ratio is defined as

$$\text{Poisson's ratio}(v) = \left| \frac{\text{lateral strain}}{\text{axial strain}} \right|$$
$$= -\frac{\text{lateral strain (contraction)}}{\text{axial strain (tension)}}$$

$$(1.9)$$

A negative sign is usually assigned to a contraction. Poisson's ratio is a dimensionless quality that is constant in the elastic range for most materials and has a value between 0 and 0.5.

1.7 GENERALISED HOOKE'S LAW

Generalised Hooke's law is an extension of the simple strain–stress relations of Equation 1.7 to a general case where stresses and strains are three dimensional.

Consider a cube subjected to normal stresses, σ_x, σ_y and σ_z, in the directions of x, y and z coordinate axes, respectively (Figure 1.12a).

From Figure 1.12, we have

Strain of case (a) = strain of case (b) + strain of case (c) + strain of case (d)

In particular, considering the normal strain of case (a) in the x direction and applying Equations 1.7 and 1.9 to cases (b), (c) and (d), we have

Normal strain in the x direction		
By σ_x Figure 1.12b	By σ_y Figure 1.12c	By σ_z Figure 1.12d
$\varepsilon_x = \dfrac{\sigma_x}{E}$	$v = -\dfrac{\varepsilon_x}{\varepsilon_y}$	$v = -\dfrac{\varepsilon_x}{\varepsilon_z}$
	$\varepsilon_x = -v\varepsilon_y$	$\varepsilon_x = -v\varepsilon_z$

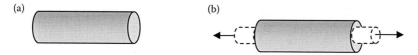

Figure 1.11 Illustration of Poisson's effect: (a) bar before loading and (b) bar after loading.

Thus, the normal strain of case (a) in the x direction is as follows:

$$\varepsilon_x = \frac{\sigma_x}{E} - v\varepsilon_y - v\varepsilon_z$$

From Figures 1.12c and d

$$\varepsilon_y = \frac{\sigma_y}{E}$$

$$\varepsilon_z = \frac{\sigma_z}{E}$$

Then

$$\varepsilon_x = \frac{\sigma_x}{E} - v\frac{\sigma_y}{E} - v\frac{\sigma_z}{E} = \frac{1}{E}[\sigma_x - v(\sigma_y + \sigma_z)] \tag{1.10}$$

The strains in the y and z directions can also be calculated by following exactly the same procedure described above. They are

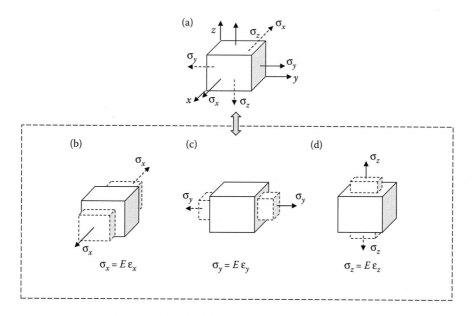

Figure 1.12 Strain the x direction caused by tri-axial stresses.

$$\varepsilon_y = \frac{\sigma_y}{E} - v\frac{\sigma_x}{E} - v\frac{\sigma_z}{E} = \frac{1}{E}[\sigma_y - v(\sigma_x + \sigma_z)]$$

$$\varepsilon_z = \frac{\sigma_z}{E} - v\frac{\sigma_x}{E} - v\frac{\sigma_y}{E} = \frac{1}{E}[\sigma_z - v(\sigma_x + \sigma_y)]$$

(1.11)

For a three-dimensional case, shear stresses and shear strains may occur within three independent planes, that is, in the x–y, x–z and y–z planes, for which the following three shear stress and strain relations exist:

$$\gamma_{xy} = \frac{\tau_{xy}}{G}$$

$$\gamma_{xz} = \frac{\tau_{xz}}{G}$$

$$\gamma_{yz} = \frac{\tau_{yz}}{G}$$

(1.12)

Equations 1.10 to 1.12 are the generalised Hooke's law. The application of Equations 1.10 to 1.12 is limited to isotropic materials in the linear elastic range.

The generalised Hooke's law of Equations 1.10 to 1.12 represents strains in terms of stresses. The following equivalent form of Hooke's law represents stresses in terms of strains:

$$\sigma_x = \frac{E(1-v)}{(1+v)(1-2v)}\left[\varepsilon_x + \frac{v}{1-v}(\varepsilon_y + \varepsilon_z)\right]$$

$$\sigma_y = \frac{E(1-v)}{(1+v)(1-2v)}\left[\varepsilon_y + \frac{v}{1-v}(\varepsilon_x + \varepsilon_z)\right]$$

$$\sigma_z = \frac{E(1-v)}{(1+v)(1-2v)}\left[\varepsilon_z + \frac{v}{1-v}(\varepsilon_x + \varepsilon_y)\right]$$

$$\tau_{xy} = G\gamma_{xy}$$

$$\tau_{xz} = G\gamma_{xz}$$

$$\tau_{yz} = G\gamma_{yz}$$

(1.13)

1.8 STRENGTH, STIFFNESS AND FAILURE

Failure is a condition that prevents a material or a structure from performing the intended task. For the cantilever shown in Figure 1.13, the following two questions, for example, can be asked:

1. What is the upper limit of stress that can be reached in the material of the beam?
   The answer to this question provides a strength criterion that can be adopted in the design of the beam (Figure 1.14):
   a. An upper limit at which the strain–stress relationship departs from linear is called the *proportional limit*, σ_{pl}.

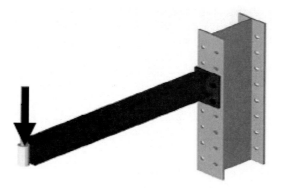

Figure 1.13 Bending of cantilever.

    b. An upper limit at which permanent or plastic deformation starts is called the *yield strength*, σ_{Yield}.

    c. An upper limit that is the maximum stress a material can withstand is called the *ultimate strength*, σ_u.

    Strength is a property of a material. In a practical design, the actual stress in a material must be smaller than a given allowable stress that is normally specified by material manufacturers or design codes. Design options include reducing the level of the design load, using a higher strength material or using more materials to restrict the stress level within a mechanical system below the safe level. Thus

$$\sigma \leq \sigma_{allowable}$$
$$\tau \leq \tau_{allowable} \tag{1.14}$$

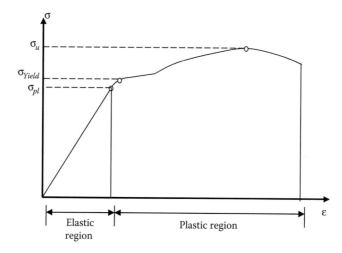

Figure 1.14 Stress–strain curve.

2. What is the maximum tip deflection that is acceptable?

The answer to this question provides a stiffness design criterion that represents the *stiffness* or the resistance of an elastic body to deformation.

Factors that influence stiffness of a structural member include material modulus, structural configuration and mode of loading. For example, the tip deflection of a cantilever varies if the materials, length, shape of cross section or the applied load change.

Strength and stiffness are measurements of resistance to failure. Violation of any of the above criteria is defined as failure. In a typical design, a primary task is to choose materials and member dimensions such that

---

- Stresses are maintained below the limits for the chosen materials.
- Deformations are maintained below the limits for the structure application.

1.9 KEY POINTS REVIEW

- An applied force can be in the form of point load, distributed load or moment.
- An applied load causes deformation and eventually failure of a structure.
- An applied force causes internal forces/stresses.
- Stress is defined as intensity of internal force at a point of material.
- A stress has magnitude and direction, and is always related to a specified plane (cross section).
- Normal stress is a stress that is perpendicular to a cross section and causes tension or compression.
- Shear stress is a stress that is parallel to a cross section and causes distortion or twisting.
- Strain is a measurement of relative deformation at a point of material, and is a non-dimensional quantity.
- Normal strain represents either an elongation or a contraction.
- Shear strain is measurement of distortion, approximately measured by change of a right angle.
- The relationship between strain and stress depends on properties of materials. For linear elastic materials, the relationship is called Hooke's law.
- For a linearly elastic and isotropic material, E, G and v are related and only two of them are independent.
- Different materials normally have different strength and strength depends on properties of material.
- Stiffness of a member depends on not only property of material, but also geometrical and loading conditions.
- Proportional limit, σ_{pl}, is the upper limit at which the strain–stress relationship departs from linear.
- Yield strength, σ_{Yield}, is the upper limit at which permanent deformation starts.
- Ultimate strength, σ_u, is the maximum stress a material can withstand.

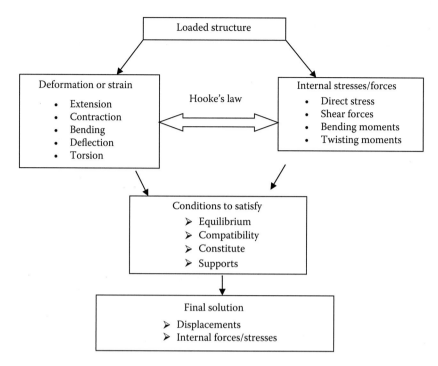

Figure 1.15 Flow chart of solution procedure.

1.10 BASIC APPROACH FOR STRUCTURAL ANALYSIS

The solution of a stress problem always follows a similar procedure that is applicable for almost all types of structures. Figure 1.15 presents a flow chart for the procedure. In general, either deformations (strains) or forces (stresses) are the quantities that need to be computed in a structural analysis of design. The following steps represent a general approach to the solution of a structural problem:

- An FBD of the entire or part of a structure is normally required.
- Calculating support reactions is usually a start point of a stress analysis. For a statically determinate structure, the reactions are determined by the application of equilibrium equations. For a statically indeterminate structure, additional equations must be sought.
- If the internal forces or stresses on a section are wanted, the method of section or/and the method of joint are used to cut through the section so that the structure is cut into two parts, that is, two free bodies.
- A part on either side of the section is taken as a free body and required to satisfy the equilibrium conditions. On the section concerned, the internal forces that keep the part in equilibrium include, in a general case, a normal force, a shear force, a bending moment and a twist moment (torque). These internal forces are found by the application of static equilibrium of all forces acting on the free body.

- Once the internal forces on the section are determined, the stresses caused by the forces can be calculated using appropriate formulas of stress analysis.
- From the stress solutions, Hooke's law can be used to compute strains and then displacements, for example, deflection of a beam subjected to bending.
- Both the stress and the strain solutions are further used in design to meet relevant strength and stiffness criterions.

1.11 EXAMPLES

EXAMPLE 1.1

Consider the rectangular bar shown in figure below. When it is subjected to uniaxial tension

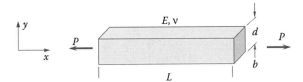

what are the normal stress and strain in the axial direction? Determine also the strain in the lateral direction.

Solution

Since the bar is subjected to uniaxial forces at the two ends only, the internal force between the two ends must be uniform. Only one section (cut) is required to take.

Step 1: Take a section between the two ends

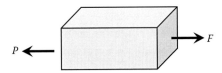

From the equilibrium in the axial direction, the internal force F is

$$F = P$$

$$\sigma_{axial} = \frac{\text{axial force}}{\text{area}} = \frac{F}{A} = \frac{P}{bd}$$

$$\varepsilon_{axial} = \frac{\sigma}{E} = \frac{P}{Ebd}$$

Step 2: Strain in the lateral direction
From Equation 1.9

$$\varepsilon_{lateral} = -v\varepsilon_{axial} = -\frac{vP}{Ebd}$$

EXAMPLE 1.2

The steel bar shown in figure below has a rectangular cross section of width 35 mm and thickness 10 mm. Determine the internal forces, normal stresses and normal strains on the cross sections between the loading points $E_{st} = 210(10^3)$ MPa

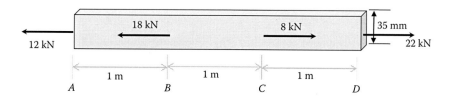

Solution

Forces act at and between the two ends, so the internal axial force along the axis is not constant and sections must be taken between any consecutive loading points. The bar has a constant cross-sectional area of $35 \times 10 = 350$ mm².

Step 1: Section taken between AB

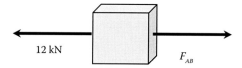

Consider equilibrium in the horizontal direction

$$F_{AB} = 12\,\text{kN}$$

$$\sigma_{AB} = \frac{F_{AB}}{A} = \frac{12\,\text{kN}}{350\,\text{mm}^2} = 34.3\,\text{MPa}$$

$$\varepsilon_{AB} = \frac{\sigma_{AB}}{E} = \frac{34.3\,\text{MPa}}{210 \times 10^3\,\text{MPa}} = 1.63 \times 10^{-4}$$

The force, stress and strain between AB are all tensile.

Step 2: Section taken between BC

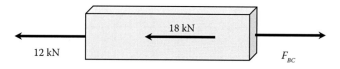

Consider equilibrium in the horizontal direction

$$F_{BC} = 12 + 18 = 30\,\text{kN}$$

$$\sigma_{BC} = \frac{F_{BC}}{A} = \frac{30\,\text{kN}}{350\,\text{mm}^2} = 85.7\,\text{MPa}$$

$$\varepsilon_{BC} = \frac{\sigma_{BC}}{E} = \frac{85.7\,\text{MPa}}{210 \times 10^3\,\text{MPa}} = 4.08 \times 10^{-4}$$

The force, stress and strain between BC are all tensile.

Step 3: Section taken between CD

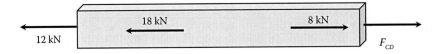

Consider equilibrium in the horizontal direction

$$F_{CD} = 12 + 18 - 8 = 22\,\text{kN}$$

$$\sigma_{CD} = \frac{F_{CD}}{A} = \frac{22\,\text{kN}}{350\,\text{mm}^2} = 62.8\,\text{MPa}$$

$$\varepsilon_{CD} = \frac{\sigma_{CD}}{E} = \frac{62.8\,\text{MPa}}{210 \times 10^3\,\text{MPa}} = 2.99 \times 10^{-4}$$

The force, stress and strain between CD are all tensile.

EXAMPLE 1.3

Calculate the support reactions and member forces of the symmetric truss shown in figure below. All the members of the truss are pin-joined. The truss is loaded with concentrated loads symmetrically applied from the top.

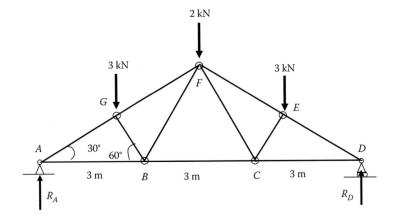

Solution

This is a statically determinate system. Both the reactions and the member forces can be found by considering equilibrium. The truss is subjected to loads applied in the vertical direction and pin-supported at A and D, which induce only vertical reactions at the supports. Since both the geometry of the truss and the loading pattern are symmetric about the central line, analysis of only half of the truss is needed.

Step 1: Calculate reaction forces at the supports

Due to symmetry, the vertical reaction forces at A and D must be the same and are equal to half of the total applied forces, that is,

$$R_A = R_D = \frac{1}{2}(3\,\text{kN} + 3\,\text{kN} + 2\,\text{kN}) = 4\,\text{kN}$$

Step 2: Calculate member forces by the method of joint

Due to symmetry, consider only joints A, B and G.

- At joint A

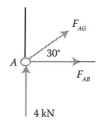

Resolving in the vertical direction

$$4 + F_{AG}\cos(60°) = 0$$

$$F_{AG} = -8\,\text{kN (compression)}$$

Resolving in the horizontal direction

$$F_{AB} + F_{AG}\cos(30°) = 0$$

$$F_{AB} = -F_{AG}\cos(30°) = -(-8)\cos(30°) = 6.93\,\text{kN (tension)}$$

- At joint G

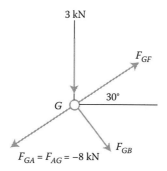

Resolving in the vertical direction

$$-3 - F_{AG}\cos(60°) - F_{GB}\cos(30°) + F_{GF}\cos(60°) = 0$$

or

$$F_{GF} - \sqrt{3}F_{GB} + 2 = 0 \tag{1.15}$$

Resolving in the horizontal direction

$$-F_{AG}\cos(30°) + F_{GB}\cos(60°) + F_{GF}\cos(30°) = 0$$

$$F_{GB} + \sqrt{3}F_{GF} + 8\sqrt{3} = 0 \tag{1.16}$$

Solving Equations 1.15 and 1.16, simultaneously,

$$F_{GB} = -2.6\,\text{kN (compression)}$$

$$F_{GF} = -6.5\,\text{kN (compression)}$$

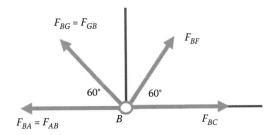

- At joint B

Resolving in the vertical direction

$$F_{BG}\cos(30°) + F_{BF}\cos(30°) = 0$$

or

$$F_{BF} = -F_{BG} = 2.6\,\text{kN (tension)}$$

Resolving in the horizontal direction

$$-F_{AB} + F_{BC} - F_{BG}\cos(60°) + F_{BF}\cos(60°) = 0$$

or

$$F_{BC} + 0.5F_{BF} - 5.63 = 0$$

Thus,

$$F_{BC} = 4.3\,\text{kN (tension)}$$

Step 3: Show member forces of all members

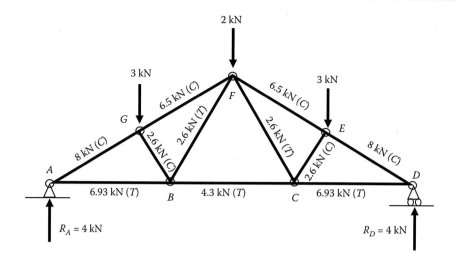

1.12 CONCEPTUAL QUESTIONS

1. What is the difference between applied loads and reactions?
2. What is meant by 'stress' and why is it a local measurement of force?
3. What is the unit for measuring stress?
4. What is the difference between a normal stress and a shear stress?
5. What is meant by 'strain'? Is it a local measure of deformation?
6. Is a 'larger strain' always related to a 'larger displacement'?
7. What is the physical meaning of 'shear strain'?
8. What is the unit for measuring strain?
9. What is Young's modulus, and how can it be determined from a simple tension test?
10. What is the simplest form of stress and strain relationship?
11. If the displacement at a point in a material is zero, the strain at the same point must be zero. Is this correct and why?
12. How is Poisson's ratio defined?
13. In a uniaxial tension test, the lateral strain is
    a. Equal to the strain in the axial direction
    b. Inversely proportional to the strain in the axial direction
    c. Proportional to the strain in the axial direction and in an opposite sense
    d. Independent to the strain in the axial direction
14. A stainless steel bar supports a uniaxial tensile stress of 100 MPa. When loaded, the axial and lateral strains of the stainless steel are, respectively, 5.26×10^{-4} and -1.58×10^{-4}. What is Poisson's ratio of the material?
    a. 0.1
    b. 0.2
    c. 0.3
    d. 0.4
15. For a linearly elastic and isotropic material, what is the relationship between Young's modulus, shear modulus and Poisson's ratio?
16. How can 'failure' of a structural member be defined?
17. What are meant, respectively, by 'proportional limit', 'yield strength' and 'ultimate strength'? Are they properties of material?
18. What is meant by 'stiffness'? Is it a property of materials and why?

19. Describe how the method of joint can be used in structural analysis.
20. Describe how the method of section can be used in structural analysis.
21. The strain–stress curve for a hypothetical material is given below. If the strain at the top and bottom of a section are, respectively, $2\varepsilon_m$ in tension and $-2\varepsilon_m$ in compression, sketch the stress distribution over the height of the section.

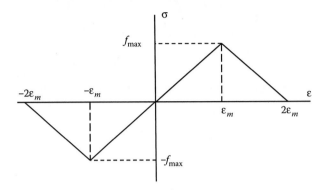

22. In a uniaxial tension test, the yield strength of a material can be defined as
   a.   The strain at which the material exceeds 5% elongation
   b.   The stress at which the material breaks
   c.   The stress at which plastic deformation starts
   d.   The stress at which the ultimate tensile stress is reached

1.13 MINI TEST

PROBLEM 1.1

Which one of the following statements is correct?

a. Normal stress on a cross section is always equal to the internal force acting on the section divided by the cross-sectional area.
b. Normal stress on a cross section is always NOT equal to the force acting on the section divided by the cross-sectional area.
c. The internal normal force acting on a cross section is the resultant of the normal stresses acting on the same section.
d. Normal stresses on a cross section are independent of the normal force acting on the same section.

PROBLEM 1.2

What are the differences between displacement, deformation and strain? Which one of the following statements is correct in relation to the loaded beam shown in figure below?

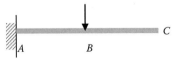

a. There are displacement, deformation and strain in BC.
b. There is only displacement in BC.
c. There are both displacement and deformation but without strain in BC.
d. There are no displacement, deformation and strain in BC.

PROBLEM 1.3

Are the longitudinal normal stress, strain and internal normal force on the cross sections of the bar (figure below) constant along its axis? Can the strain of the bar be calculated by $\varepsilon = \Delta L/L$ and why?

If the bar has a circular section whose largest and smallest diameters are D and d, respectively, calculate the strain along the bar. Assume that Young's modulus of the materials is E.

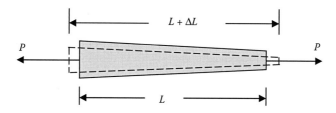

PROBLEM 1.4

Use the method of section to determine the internal forces on the cross section at B (figure below).

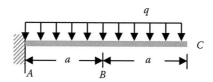

PROBLEM 1.5

Use the method of joint to determine the axial forces in the members of the truss shown in figure below.

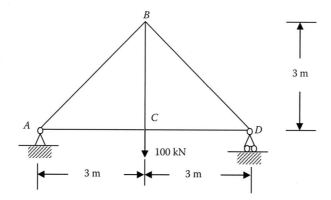

Chapter 2

Axial tension and compression

In practical situations, axial tension or compression is probably the simplest form of deformation. This type of deformation is characterised by the following:

- The action line of the resultant of applied forces coincides with the axis of the member.
- Under the axial force, normal stress develops on cross sections.
- Under the axial force, the deformation of the member is dominated by either axial elongation or axial shortening with associated contraction or expansion, respectively, in the lateral direction.

2.1 SIGN CONVENTION

A positive axial force (stress) is defined as a force (stress) that induces elongation (Figure 2.1a). A negative axial force (stress) is defined as a force (stress) that induces axial shortening (Figure 2.1b).

The sign convention is designed to characterise the nature of the force or stress, rather than in relation to a particular direction of the coordinates. For example, in Figure 2.1a, both forces are positive because they are all tensile. While setting up equilibrium equation, the two forces are opposite, that is, one is positive and the other is negative.

2.2 NORMAL (DIRECT) STRESS

The uniformly distributed normal stress, σ, on section m is calculated by

$$\sigma = \frac{F}{A} \tag{2.1}$$

where A is the cross-sectional area of the bar; σ takes the sign of F (Figure 2.2). Since the force and the cross-sectional area are both constant along the bar in this case, the normal stress is also constant along the bar. This is not always true if either the force or the cross-sectional area is variable.

Applying the simple form of Hooke's law (Equation 1.5) to the bars yields

$$\varepsilon = \frac{\sigma}{E} = \frac{F}{EA} \tag{2.2}$$

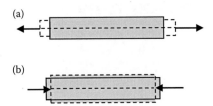

Figure 2.1 Bar subjected to uniaxial force: (a) tension and (b) compression.

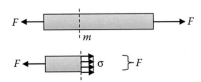

Figure 2.2 Normal stress on cross section.

Again this applies only when both the internal normal force and the cross-sectional area are constant along the bar.

2.3 STRESSES ON AN ARBITRARILY INCLINED PLANE

There are situations in which stresses on a plane that is not perpendicular to the member axis are of interest, for example, the direct stress σ_α and shear stress τ_α along the interface of the adhesively bonded scarf joint shown in Figure 2.3.

From the equilibrium of

$$\sigma = \frac{F}{A}$$

$$\text{Normal stress (peeling stress) } \sigma_\alpha = \frac{F}{A}\cos^2\alpha = \sigma\cos^2\alpha \qquad (2.3)$$

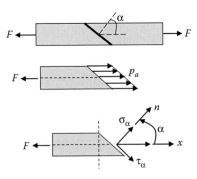

Figure 2.3 Stress on an inclined plane.

$$\text{Shear stress } \tau_\alpha = \frac{F}{2A}\sin 2\alpha = \frac{\sigma}{2}\sin 2\alpha \qquad (2.4)$$

$$\text{Resultant stress } p_a = \frac{F}{A}\cos \alpha = \sigma \cos \alpha \qquad (2.5)$$

Important observations:

> a. The maximum normal stress occurs when $\alpha = 0$, that is, $\sigma_{max} = \sigma$.
> b. The maximum shear stress occurs when $\alpha = \pm 45°$, that is, $\tau_{max} = \sigma/2$.
> c. On the cross section where maximum normal stress occurs ($\alpha = 0$), there is no shear stress.

2.4 DEFORMATION OF AXIALLY LOADED MEMBERS

2.4.1 Members of uniform sections

Equations 1.1, 1.2, 1.7, 2.1 and 2.2 are sufficient to determine the deformation of an axially loaded member.

Axial deformation (see Figure 2.4)

$$\Delta l = l_1 - l = \frac{Nl}{EA} \qquad (2.6)$$

where N is the internal axial force due to the action of F ($N = F$ in Figure 2.4).

When the internal axial force or/and the cross-sectional area vary along the axial direction

$$\Delta l = \int_l \frac{N(x)}{EA(x)}\,dx \qquad (2.7)$$

2.4.2 Member with step changes

An axially loaded bar may be composed of segments with different cross sections, different materials and be loaded between the two ends, as shown in Figure 2.5.

The internal axial force, and therefore the normal stress on the cross sections, will not be constant along the bar (except Figure 2.5c, where normal strain is not constant). They are constant within each segment, of which cross-sectional areas, internal normal stresses and

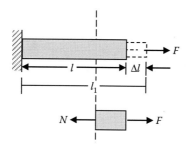

Figure 2.4 Elongation of an axially loaded bar.

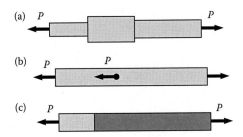

Figure 2.5　Varying stress or strain along the bar.

materials are all constant. Thus, to determine the stress, strain and elongation of such a bar, each segment must be considered independently.

2.5 STATICALLY INDETERMINATE AXIAL DEFORMATION

There are axially loaded structural members whose internal forces cannot be determined simply by the equations of equilibrium. This type of structures is called *statically indeterminate structure*. For the bar shown in Figure 2.6, for example, the two reaction forces at the fixed ends cannot be uniquely determined by considering only the equilibrium in the horizontal direction. The equilibrium condition gives only a single equation in terms of the two unknown reaction forces, R_A and R_B. An additional condition must be sought in order to form the second equation. Usually, for a statically indeterminate structure, additional equations come from considering deformation of the system.

In Figure 2.6, we can see that the overall elongation of the bar is zero since it is fixed at both ends. This is called *geometric compatibility of deformation*, which provides an additional equation for the solution of this problem. The total deformation of the bar is calculated considering the combined action of the two unknown support reactions and the externally applied loads. Example 2.3 shows how this condition can be used to form an equation in terms of the two reaction forces. In general, the following procedure can be adapted:

- Replace supports by reaction forces.
- Establish static equilibrium equation.
- Consider deformation of members to form a geometric relationship (equation).
- Use the force–deformation relationships, for example, Equation 2.6 for axial deformation, and introduce them to the geometric equation.
- Solve a simultaneous equation system that consists of both the static equilibrium equations and the geometric compatibility equations.

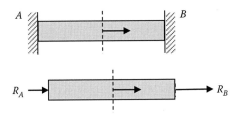

Figure 2.6　Deformation of a statically indeterminate bar.

2.6 ELASTIC STRAIN ENERGY OF AN AXIALLY LOADED MEMBER

Strain energy is the internal work done in a body by externally applied forces. It is also called *the internal elastic energy of deformation.*

2.6.1 Strain energy U in an axially loaded member

For an axially loaded member with constant internal axial force

$$U = \frac{N\Delta l}{2} = \frac{N^2 l}{2EA} \tag{2.8}$$

If the member is composed of segments with different cross-sectional areas, A_i, internal normal forces, N_i, and materials, E_i, the strain energy is the sum of the energy stored in each of the segments as

$$U = \sum_{i=1}^{n} \frac{N_i \Delta l_i}{2} = \sum_{i=1}^{n} \frac{N_i^2 l_i}{2E_i A_i} \tag{2.9}$$

2.6.2 Strain energy density, U_0

Strain energy density is the strain energy per unit volume. For the axially loaded member with constant stress and strain

$$\begin{aligned} U_0 &= \frac{U}{\text{volume}} = \frac{N^2 l/2EA}{Al} \\ &= \frac{\sigma\varepsilon}{2} = \frac{E\varepsilon^2}{2} = \frac{\sigma^2}{2E} \end{aligned} \tag{2.10}$$

U_0 is usually measured in J/m^3.

2.7 SAINT-VENANT'S PRINCIPLE AND STRESS CONCENTRATION

Equations 2.1 and 2.2 assume that normal stress is constant across the cross section of a bar under an axial load. However, the application of this formula has certain limitations.

When an axial load is applied to a bar of a uniform cross section as shown in Figure 2.7a, the normal stress on the sections away from the localities of the applied concentrated loads will be uniformly distributed over the cross section. The normal stresses on the sections near the two ends, however, will not be uniformly distributed because of the non-uniform deformation caused by the applied concentrated loads. Obviously, a higher level of strain, and hence, higher level of stress, will be induced in the vicinity of the applied loads.

This observation is often useful when solving static equilibrium problems. It suggests that if the distribution of an external load is altered to a new distribution that is however *statically equivalent* to the original one, the stress distribution on a section sufficiently far from where the alteration was made will be little affected. This conclusion is termed as *Saint-Venant's principle.*

The unevenly distributed stress in a bar can also be observed if a hole is drilled in a material (Figure 2.7b). The stress distribution is not uniform on the cross sections near the

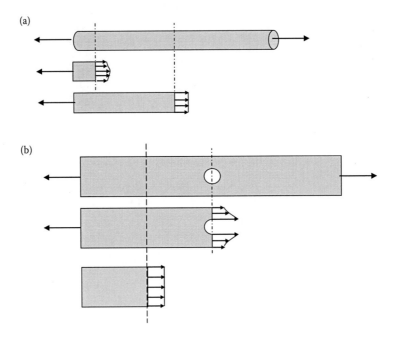

Figure 2.7 Illustration of Saint-Venant's principle and stress concentration.

hole. Since the material that has been removed from the hole is no longer available to carry any load, the load must be redistributed over the remaining material. It is not redistributed evenly over the entire remaining cross-sectional area, with a higher level of stress near the hole. On the cross sections away from the hole, the normal stress distribution is still uniform. The unevenly distributed stress is called *stress concentration*.

This observation suggests that if a structural member has a sudden change of cross-section shape or discontinuity of geometry, stress concentration will occur in the vicinity of the sudden changes or discontinuities.

2.8 STRESS CAUSED BY TEMPERATURE

In a statically determinate structure, the deformation due to temperature changes is usually disregarded, since in such a structure the members are free to expand or contract. However, in a statically indeterminate structure, expansion or contraction can be restricted. This can sometimes generate significant stresses that may cause failure of a member and eventually of the entire structure.

For a bar of length L (Figure 2.8), the free deformation caused by a change in temperature, ΔT, is

$$\Delta L_T = \alpha_T \Delta T L \tag{2.11}$$

and the thermal strain is, therefore,

$$\varepsilon_T = \alpha_T \Delta T \tag{2.12}$$

where α_T is the coefficient of thermal expansion.

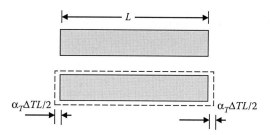

Figure 2.8 Thermal expansion of a material.

If the bar is not completely free to expand or contract, the strain of the bar is a sum of the above thermal strain and the strain caused by the stress that is developed due to the restraints to free expansion, that is,

$$\varepsilon = \frac{\sigma}{E} + \alpha_T \Delta T \qquad (2.13)$$

2.9 KEY POINTS REVIEW

- A shaft/rod is long compared to its other two dimensions.
- The load applied is along the axial direction.
- If a shaft is subjected to axially applied loads, the deformation can be defined by either axial tension or axial compression.
- The cross section deforms uniformly.
- If a structural member is free to expand and contract, a change in temperature will generate strains, but not stresses.
- If a structural member is prevented to expand and contract, a change in temperature will generate both strains and stresses.
- The internal stresses and forces on sections perpendicular to the axis are normal stresses and axial forces, respectively.
- For an axially loaded bar, the maximum normal stress at a point is the normal stress on the section perpendicular to the axis.
- For an axially loaded bar, the maximum shear stress at a point is the shear stress acting on the plane that is 45° to the axis.
- On the cross section where maximum normal stress occurs, there exists no shear stress.
- On the plane of maximum shear stress, the normal stress is not necessarily zero.
- The resultant of normal stress on a cross section is the axial force acting on the same section.
- The axial tension or compression stiffness of a section is *EA*.
- Strain energy is proportional to square of forces, stresses or strains.
- For a structure system, if the number of independent equilibrium equations is less than the number of unknown forces, the system is termed as an *indeterminate system*.

- To solve a statically indeterminate system, geometrical compatibility of deformation must be considered along with the equilibrium conditions.
- Replacing a load applied on a material by an alternative, but *statically equivalent* load will affect only the stress field in the vicinity.
- An abrupt change of geometry of a structural member will cause stress concentration near the region of the change.

2.10 RECOMMENDED PROCEDURE OF SOLUTION

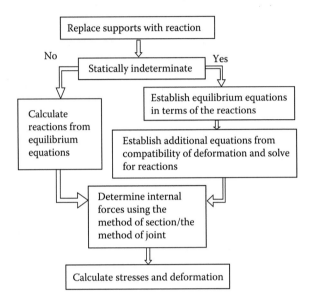

2.11 EXAMPLES

EXAMPLE 2.1

The uniform bar is loaded as shown in figure (a) below. Determine the axial stress along the bar and the total change in length. The cross-sectional area and Young's modulus of the bar are, respectively, $A = 1 \text{ cm}^2$ and $E = 140$ GPa.

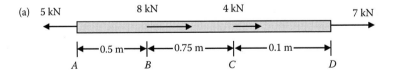

Solution

This is a uniform bar with forces applied between the two ends. The internal forces along the bar are, therefore, not constant. The stresses and elongation between A and B, B and C, and C and D must be calculated individually and the total change in length is equal to the sum of all the elongations. The method of section is used to determine the internal forces.

Between A and B

(b)

From the equilibrium of figure (b) above

$N_{BC} - 5\,\text{kN} = 0$

$N_{BC} = 5\,\text{kN}$

The normal stress between A and B (Equations 2.1 and 2.2)

$$\sigma_{AB} = \frac{N_{AB}}{A} = \frac{5\,\text{kN}}{0.0001\,\text{m}^2} = 5 \times 10^4\,\text{kN/m}^2 = 50\,\text{MPa}$$

The elongation between A and B (Equation 2.6)

$$\Delta_{AB} = \frac{N_{AB}L_{AB}}{EA} = \frac{5\,\text{kN} \times 0.5\,\text{m}}{140 \times 10^6\,\text{kN/m}^2 \times 0.0001\,\text{m}^2} = 178.57 \times 10^{-6}\,\text{m}$$

Between B and C

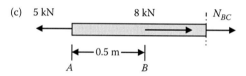

From the equilibrium of figure (c) above

$N_{AB} + 8\,\text{kN} - 5\,\text{kN} = 0$

$N_{AB} = -8\,\text{kN} + 5\,\text{kN} = -3\,\text{kN}$ (compression)

The normal force should be in the opposite direction of the assumed N_{BC}.
The normal stress between B and C (Equation 2.1)

$$\sigma_{BC} = \frac{N_{BC}}{A} = \frac{-3\,\text{kN}}{0.0001\,\text{m}^2} = -3 \times 10^4\,\text{kN/m}^2 = -30\,\text{MPa}$$

The elongation between B and C (Equation 2.6)

$$\Delta_{BC} = \frac{N_{BC}L_{BC}}{EA} = \frac{-3\,\text{kN} \times 0.75\,\text{m}}{140 \times 10^6\,\text{kN/m}^2 \times 0.0001\,\text{m}^2} = -160.71 \times 10^{-6}\,\text{m}$$

Between C and D

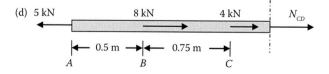

From the equilibrium of figure (d) above

$$N_{CD} + 8\,\text{kN} + 4\,\text{kN} - 5\,\text{kN} = 0$$

$$N_{CD} = -8\,\text{kN} - 4\,\text{kN} + 5\,\text{kN} = -7\,\text{kN (compression)}$$

The normal force should be in the opposite direction of the assumed N_{CD}.
The normal stress between C and D (Equation 2.1)

$$\sigma_{CD} = \frac{N_{CD}}{A} = \frac{-7\,\text{kN}}{0.0007\,\text{m}^2} = -7 \times 10^4\,\text{kN/m}^2 = -70\,\text{MPa}$$

The elongation between C and D (Equation 2.6)

$$\Delta_{CD} = \frac{N_{CD}L_{CD}}{EA} = \frac{-7\,\text{kN} \times 1.0\,\text{m}}{140 \times 10^6\,\text{kN/m}^2 \times 0.0001\,\text{m}^2} = -500 \times 10^{-6}\,\text{m}$$

The total change in length of the bar is

$$\Delta_{\text{Total}} = \Delta_{AB} + \Delta_{BC} + \Delta_{CD} = 178.57 \times 10^{-6}\,\text{m} - 160.71 \times 10^{-6}\,\text{m} - 500 \times 10^{-6}\,\text{m}$$

$$= -482.14 \times 10^{-6}\,\text{m} = -0.482\,\text{mm}$$

The negative sign means the change in length is contraction.

EXAMPLE 2.2

The concrete pier shown in figure (a, b) below supports a uniform pressure of 20 kN/m^2 at the top. The density of the concrete is 25 kN/m^3 and the pier is 0.5 m thick. Calculate the reaction force at the base and the stress at a level of 1 m above the base.

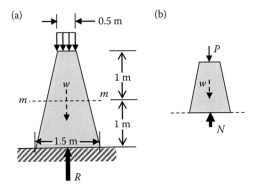

Solution

This is a statically determinate structure subjected to only vertical loads. The vertical reaction at the base can be determined from vertical equilibrium. The method of section can be used to calculate the stress on section m–m.

From the equilibrium in the vertical direction, the reaction R must be equal to the sum of the weight and the resultant of the pressure.

Weight of the pier

$$W = \frac{(0.5 + 1.5) \times 2}{2} \times 0.5 \times 25 = 25\,\text{kN}$$

Resultant of the pressure

$$P = 0.5 \times 0.5 \times 20 = 5\,\text{kN}$$

Thus, the reaction at base

$$R = W + P = 30\,\text{kN}$$

Cut at *m–m* and take the upper part of the pier as a free body as shown in figure (b) above. Forces from the part above the cut

$$P + W = 5 + \frac{(0.5 + 1) \times 1}{2} \times 0.5 \times 25 = 14.4\,\text{kN}$$

From the vertical equilibrium of figure (b) above

$$N = P + W = 14.4\,\text{kN}$$

Thus, the stress at this level is

$$\sigma = \frac{N}{A} = \frac{14.4\,\text{kN}}{1.5\,\text{m} \times 0.5\,\text{m}} = 19.2\,\text{kN/m}^2 \text{ (compression)}$$

EXAMPLE 2.3

A bar made of steel and brass has the dimensions shown in figure (a) below. The bar is rigidly fixed at both ends. Calculate the end reactions and stresses when the force F is applied at level C. Let $A_{steel} = 2 \times 10^4\,\text{mm}^2$, $E_{steel} = 200$ GPa, $A_{br} = 1 \times 10^4\,\text{mm}^2$, $E_{br} = 100$ GPa and $F = 1 \times 10^3$ kN.

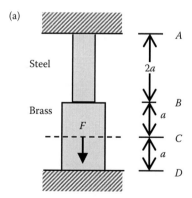

Solution

This is a statically indeterminate system subjected to only vertical forces. The equilibrium in the vertical direction must be first established. Due to the two fixed ends, the total elongation of the bar is zero. This observation provides the additional equation in terms of the two unknown reactions.

(b)

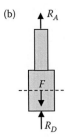

From the vertical equilibrium of figure (b) above

$$R_A + R_D - F = 0 \qquad (2.14)$$

Since the total deformation between A and D is zero

$$\Delta L_{AB} + \Delta L_{BC} - \Delta L_{CD} = 0$$

where ΔL_{AB}, ΔL_{BC} and ΔL_{CD} are the respective deformations between A and B, B and C, and C and D. The negative sign before ΔL_{CD} denotes compressive deformation between C and D. From Equation 2.6

$$\Delta L_{AB} = \frac{R_A \times 2a}{E_{steel} A_{steel}}$$

$$\Delta L_{BC} = \frac{R_A a}{E_{br} A_{br}}$$

$$\Delta L_{CD} = \frac{R_D a}{E_{br} A_{br}}$$

Thus,

$$\frac{R_A \times 2a}{E_{steel} A_{steel}} + \frac{R_A a}{E_{br} A_{br}} - \frac{R_D a}{E_{br} A_{br}} = 0$$

or

$$\frac{2R_A}{200 \times 10^9 \, \text{Pa} \times 100 \times 10^{-4} \, \text{m}^2} + \frac{R_A - R_D}{100 \times 10^9 \, \text{Pa} \times 200 \times 10^{-4} \, \text{m}^2} = 0$$

or

$$3R_A - R_D = 0 \qquad (2.15)$$

From Equations 2.14 and 2.15

$$R_D = \frac{3}{4}F = 750 \, \text{kN}, \quad R_A = \frac{1}{4}F = 250 \, \text{kN}$$

Stresses in the bar

Stress between A and B $\sigma_{AB} = \dfrac{R_A}{A_{steel}} = \dfrac{250 \, \text{kN}}{100 \times 10^{-4} \, \text{m}^2} = 25 \, \text{MPa (tension)}$

Stress between B and C $\sigma_{BC} = \dfrac{R_A}{A_{br}} = \dfrac{250\,\text{kN}}{200 \times 10^{-4}\,\text{m}^2} = 12.5\,\text{MPa (tension)}$

Stress between C and D $\sigma_{CD} = \dfrac{R_D}{A_{br}} = \dfrac{750\,\text{kN}}{200 \times 10^{-4}\,\text{m}^2} = 37.5\,\text{MPa (compression)}$

EXAMPLE 2.4

A rigid beam of $3a$ long is hinged at one end and supported by two steel wires as shown in figure (a–c) below. Wire 1 is 0.1 mm short due to a manufacturing error and has to be stretched so as to be connected to the beam. If the ratio between the cross-sectional areas of the two wires, A_1/A_2, is 2 and the allowable stress of steel is 160 MPa, calculate the minimum cross-sectional areas of both wires. $E_{steel} = 200$ GPa.

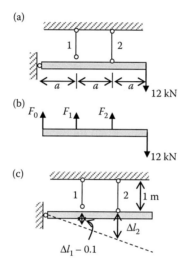

Solution

This is a statically indeterminate structure. Apart from the equilibrium conditions, the relationship between the elongation of wires 1 and 2 can be used as the compatibility condition. Note that the beam can only rigidly rotate about the pin.

From the equilibrium of moment about the pinned end (figure [b] above)

$$F_1 a + 2F_2 a - 12 \times 3a = 0 \tag{2.16}$$

From the deformation shown in figure (c) above

$$\Delta L_2 = 2(\Delta L_1 - 0.1)$$

where ΔL_1 and ΔL_2 represent, respectively, the elongation of wires 1 and 2. Thus, from Equation 2.6

$$\Delta L_1 = \frac{F_1 L_1}{E_{steel} A_1}$$

$$\Delta L_2 = \frac{F_2 L_2}{E_{steel} A_2}$$

then

$$\frac{F_2 L_2}{E_{steel} A_2} = 2\left(\frac{F_1 L_1}{E_{steel} A_1} - 0.1\right)$$

or

$$\frac{F_2 \times 1 \times 10^3}{200 A_2} = 2\left(\frac{F_1 \times 999}{2 \times 200 \times A_2} - 0.1\right) \qquad (2.17)$$

In Equation 2.17 A_1 is replaced by $2A_2$ and kN and mm are used as the basic units. Solving Equations 2.16 and 2.17 simultaneously yields

$$F_1 = 12 + 0.0267 A_2$$
$$F_2 = 12 - 0.0267 A_2$$

To meet the design requirement

$$\frac{F_1}{A_1} = \frac{12 + 0.0267 A_2}{A_1} = \frac{12 + 0.0267 A_2}{2A_2} \leq 160 \times 10^{-3}\,\text{kN/mm}^2 \qquad (2.18)$$

$$\frac{F_2}{A_2} = \frac{12 - 0.0267 A_2}{A_2} = \frac{12 - 0.0133 A_2}{A_2} \leq 160 \times 10^{-3}\,\text{kN/mm}^2 \qquad (2.19)$$

From Equation 2.18

$$A_2 = 40.91\,\text{mm}^2$$

From Equation 2.19

$$A_2 = 69.24\,\text{mm}^2$$

To meet both requirements, $A_2 = 69.24$ mm² and $A_1 = 2A_2 = 138.48$ mm² are the respective minimum areas of the supporting wires.

EXAMPLE 2.5

A concrete cylinder of area A_c is reinforced concentrically with a steel bar of area A_{steel} (figure [a] below). The composite unit is L long and subjected to a uniform temperature change of ΔT. Compute the thermal stresses in the concrete and steel. The thermal coefficients and Young's modulus of the concrete and steel are, respectively, α_c and α_{steel} and E_c and E_{steel}.

(a)

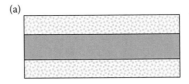

Solution

Due to the difference in the coefficients of thermal expansion, the incompatibility of the longitudinal thermal deformation between the concrete and steel will cause internal stresses. When the concrete and steel work as a unit, the final expansion of the cylinder, indicated by L_{final} in figure (b) below, is the result of both free thermal expansion, ΔL_c^{free} and ΔL_{steel}^{free}, and the deformation due to the stresses induced by the incompatibility Δ_{N_c} and $\Delta_{N_{steel}}$. For both concrete and steel, the relationships between these deformations need to be established for the solution. Because this is a symmetric system, the middle section has no displacement. Thus, only half length of the composite cylinder is considered.

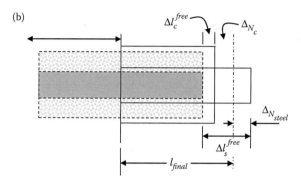

(b)

1. Thermal expansion of the concrete and steel if they are not bonded together
   For the concrete, from Equation 2.9

$$\Delta L_c^{free} = \alpha_c \Delta T \frac{L}{2}$$

For the steel

$$\Delta L_{steel}^{free} = \alpha_{steel} \Delta T \frac{L}{2}$$

2. Deformation due to thermal stresses
   Assume that the stresses on the cross sections of both the concrete and steel are uniform. The resultants of these stresses in concrete and steel are, respectively, N_c and N_{steel}. Thus, the deformation due to these forces is as follows:
   For the concrete

$$\Delta_{N_c} = \frac{N_c L/2}{E_c A_c}$$

For the steel

$$\Delta_{N_{steel}} = \frac{N_{steel} L/2}{E_{steel} A_{steel}}$$

3. Geometric relation between the above deformations
   From figure (b) above, l_{final} is the final position of the composite cylinder. Thus

$$\Delta_{N_c} + \Delta_{N_{steel}} = \Delta L_{steel}^{free} - \Delta L_c^{free}$$

or

$$\frac{N_c L/2}{E_c A_c} + \frac{N_{steel} L/2}{E_{steel} A_{steel}} = \alpha_{steel} \Delta T \frac{L}{2} - \alpha_c \Delta T \frac{L}{2} = (\alpha_{steel} - \alpha_c) \Delta T \frac{L}{2}$$

From the equilibrium of the composite cylinder

$$N_c = N_{steel}$$

Thus,

$$N_c = N_{steel} = \frac{E_{steel} A_{steel} E_c A_c}{E_{steel} A_{steel} + E_c A_c} (\alpha_{steel} - \alpha_c)\Delta T$$

The stresses in the concrete and steel are therefore

$$\sigma_c = \frac{N_c}{A_c} = \frac{E_{steel} A_{steel} E_c}{E_{steel} A_{steel} + E_c A_c} (\alpha_{steel} - \alpha_c)\Delta T$$

$$\sigma_{steel} = \frac{N_{steel}}{A_{steel}} = \frac{E_{steel} E_c A_c}{E_{steel} A_{steel} + E_c A_c} (\alpha_{steel} - \alpha_c)\Delta T$$

2.12 CONCEPTUAL QUESTIONS

1. Explain the terms 'stress' and 'strain' as applied to an elastic bar in tension. What is the relationship between the two quantities?
2. What is meant by 'elastic material'? Define modulus of elasticity.
3. Why does the axial stress vary along an axially loaded bar of variable section?
4. Consider the stresses in an axially loaded member. Which of the following statements are correct?
   a. On the section where the maximum tensile stress occurs, shear stress vanishes.
   b. On the section where the maximum tensile stress occurs, shear stress exists.
   c. On the section where the maximum shear stress occurs, normal stress vanishes.
   d. On the section where the maximum shear stress occurs, normal stress may exist.
5. A section subjected to maximum axial force is the section that will fail first. Is this correct, and why?
6. Two bars that are geometrically identical are subjected to the same axial loads. The bars are made of steel and timber, respectively. Which of the following statements is correct? The bars have
   a. The same stresses and strains
   b. Different stresses and strains
   c. The same stresses but different strains
   d. The same strains but different stresses
7. Which following statements are correct about elastic modulus E?
   a. E represents the capacity of a material's resistance to deformation.
   b. E is a material constant and is independent of stress level in a material.
   c. E depends on cross-sectional area when a member is in tension.
   d. E depends on the magnitude and direction of the applied load.
8. What is your understanding of Saint-Venant's principle? What is the importance of the principle in relation to static stress analysis?
9. Two identical bars are subjected to axial tension as shown in figure below. If the normal stresses on the cross section at the middle of both bars are the same.
   a. What is the relationship between the concentrated force P and the uniformly distributed end pressure p?
   b. Is the distribution of normal stress on the sections near the ends of the bars the same as the distribution on the midsection and why?

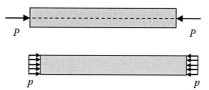

10. What is meant by 'stress concentration'? Which of the bars shown in figure below is more sensitive to stress concentration?

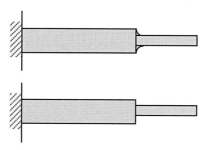

11. Three bars are made of the same material and the shapes of their cross section are, respectively, solid square, solid circle and hollow circle. If the bars are subjected to the same axial force, which of the following statements is correct?
    a. To achieve the same level of stiffness, the square section uses less material.
    b. To achieve the same level of stiffness, the circular section uses less material.
    c. To achieve the same level of stiffness, the hollow circular section uses less material.
    d. To achieve the same level of stiffness, the three sections use the same amount of material.

12. The symmetric space truss consists of four members of equal length (figure below). The relationship between the axial stiffness of the members is $E_1A_1 > E_2A_2 > E_3A_3 > E_4A_4$. If a vertical force F is applied at the joint of the four members, in which member does the maximum axial force occur and why?

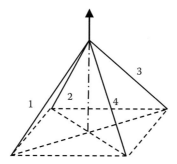

13. The following members are subjected to a uniform change in temperature. In which members will the temperature change cause stresses in the axial direction?

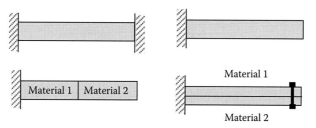

2.13 MINI TEST

PROBLEM 2.1

The four bars shown in figure below are made of different materials and have an identical cross-sectional area A. Can the normal stresses on the sections in middle span be calculated by P/A, and why?

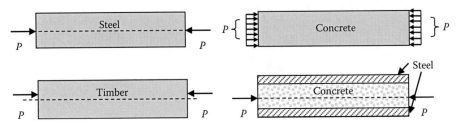

PROBLEM 2.2

The bar with variable section is subjected to an axial force P as shown in figure below.

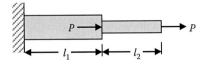

The total elongation, the strain and the total strain energy of the bar are calculated as follows:

Total elongation	$\Delta l = \Delta l_1 + \Delta l_2 = \dfrac{2Pl_1}{EA_1} + \dfrac{Pl_2}{EA_2}$
Strain	$\varepsilon = \dfrac{\Delta l_1}{l_1} + \dfrac{\Delta l_2}{l_2}$
Total strain energy	$U = U_1 + U_2 = \dfrac{2P^2 l_1}{2EA_1} + \dfrac{P^2 l_2}{2EA_2}$

Are the calculations correct, and why?

PROBLEM 2.3

A force of 500 kN is applied to the pin-joined truss as shown in figure below. Determine the required cross-sectional area of the members if the allowable stresses are 100 MPa in tension and 70 MPa in compression.

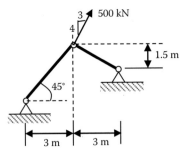

PROBLEM 2.4

The members of the pin-joined truss shown in figure below have the same tensional stiffness EA. Determine the axial forces developed in the three members due to the applied force P.

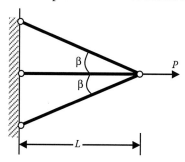

PROBLEM 2.5

A weight of 30 kN is supported by a short concrete column of square section 250×250 mm². The column is strengthened by four steel bars in the corners of total cross-sectional area 6×10^3 mm². If the modulus of elasticity for steel is 15 times that for concrete, find the stresses in the steel and the concrete.

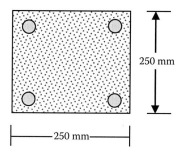

250 mm

250 mm

If the stress in the concrete must not exceed 400 kN/m², what area of steel is required in order that the column may support a load of 60 kN?

Chapter 3

Torsion

Torsion is a basic type of deformation of a structural member when it is subjected to a twist action of applied forces, as shown by the cantilever shaft subjected to a torque at the free end (Figure 3.1). If the shaft is long and has a circular section, its torsion and deformation are characterised by the following:

- The torque or twist moment is applied within a plane perpendicular to the axis of the circular member.
- Under the action of the torque, shear stress develops on the cross sections.
- Under the action of the torque, the deformation of the bar is dominated by angle of twist, that is, the relative rotation between parallel planes perpendicular to the axis.
- If angle of twist is small, the length of shaft and its radius remain unchanged.
- A plane section perpendicular to the axis remains plane after the twist moment is applied, that is, no warpage or distortion of parallel planes normal to the axis of a member occurs.
- In a circular member subjected to torsion, both shear stresses and shear strains vary linearly from the central axis.

3.1 SIGN CONVENTION

A positive torque is a moment that acts on the cross section in a right-hand-rule sense about the outer normal to the cross section (Figure 3.2), where torque is defined as positive when thumb is directed outward from the shaft. Consequently, a positive angle of twist is a rotation of the cross section in a right-hand-rule sense about the outer normal. As defined from the sign convention for the normal forces in Chapter 2, the sign convention defined here is again not related to a particular direction of coordinates.

3.2 SHEAR STRESS

Stresses that are developed on a cross section due to torsion are parallel to the section and, therefore, are shear stresses. In a circular member subjected to a torque, shear strain, γ, varies linearly from zero at the central axis. By Hooke's law (Equation 1.7), shear stress, τ, is proportional to shear strain, that is,

$$\tau = G\gamma \qquad (3.1)$$

where G denotes shear modulus of material.

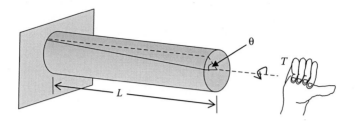

Figure 3.1 Torsional deformation of a circular bar.

On a cross section, shear stress also varies linearly from the central axis (Figure 3.3)

$$\tau = \frac{Tr}{J} \qquad (3.2)$$

where

T = torque acting on the section
r = radial distance from the centre
J = polar second moment of area of section, representing a geometric quantity of the cross section and having a unit of, for example, m⁴. The mathematical expression of J is

$$J = \int_{area} r^2 dA(\text{rea}) \qquad (3.3)$$

For the tubular section shown in Figure 3.3b

$$J = \frac{\pi}{32}(D_{out}^4 - D_{in}^4) \qquad (3.4)$$

where D_{out} and D_{in} are, respectively, the outside and inside diameters of the section. For the solid section shown in Figure 3.3a, the inner diameter (D_{in}) equals zero, that is,

$$J = \frac{\pi}{32}D_{out}^4 \qquad (3.5)$$

3.3 ANGLE OF TWIST

Angle of twist is the angle difference between two parallel sections of a bar subjected to torsion. It is proportional to the applied torsion, T, and the distance between the two sections, L,

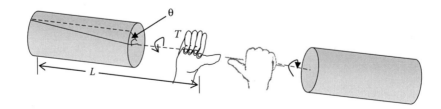

Figure 3.2 Sign convention.

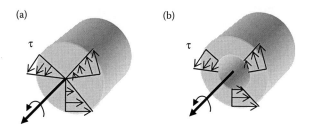

Figure 3.3 Shear stress distribution on cross section.

while inversely proportional to the geometric quantity of cross section, J, and the shear modulus, G, where GJ is called torsional rigidity. Thus, for the bar shown in Figure 3.4

$$\theta = \frac{TL}{GJ} \tag{3.6}$$

If the changes in twist moment, cross-sectional geometry and shear modulus along the central axis between sections are discrete, the total angle of twist is

$$\theta = \sum_{i=1}^{N} \frac{T_i L_i}{G_i J_i} \tag{3.7}$$

where N is the total number of the discrete segments ($i = 1, 2, ..., N$), within each of which T_i, G_i and J_i are all constant.

If the changes are continuous for T, J and G within length L

$$\theta = \int_0^L \frac{T(x)}{G(x)J(x)} \, dx \tag{3.8}$$

3.4 TORSION OF ROTATING SHAFTS

Members as rotation shafts for transmitting power are usually subjected to torques. The following formula is used for the conversion of kilowatts (kW), a common unit used in the industry, into torque applied on a shaft:

$$T = 159 \frac{p}{f}$$

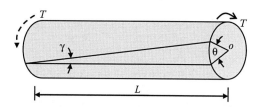

Figure 3.4 Angle of rotation between two sections.

or

$$T = 9540 \frac{p}{\text{Rev}} \tag{3.9}$$

where

T = torque (Nm)

p = transmitted power (kW)

f = frequency of rotating shaft (Hz)

Rev = revolutions per minutes of rotating shaft (rpm)

3.5 KEY POINTS REVIEW

- If a shaft is long and the load applied is twist moments/torques about the longitudinal axis, the deformation can be defined by the angle of twist.
- The internal stresses and forces on internal cross sections perpendicular to the axis are shear stresses and torques, respectively.
- The resultant of the shear stresses on any cross section along the shaft is the torque acting on the same section.
- The polar second moment of area and the shear modulus represent, respectively, the geometric and material contributions of the shaft to the torsional stiffness.
- The torsional stiffness of a section is GJ, which defines torsional resistance of a member.
- Shear stress is proportional to shear strain and varies linearly from zero at the central axis.
- On a cross section, maximum shear stress always occurs at a point along the outside boundary of the section.
- Equations 3.2 to 3.8 do not apply for noncircular members, for which cross sections perpendicular to the axis warp when a torque is applied.

3.6 RECOMMENDED PROCEDURE OF SOLUTION

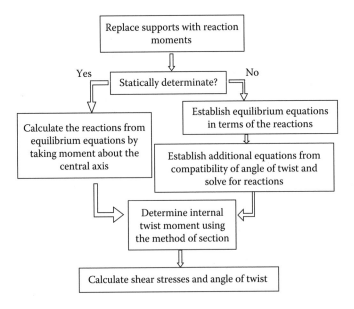

3.7 EXAMPLES

EXAMPLE 3.1

The solid shaft shown in figure below is subjected to two concentrated torques, respectively, at A and B. The diameter of the shaft changes from 1 to 0.5 m at B. What is the total angle of twist at locations A and B and what is the maximum shear stress within the shaft? $E = 200$ GPa and $v = 0.25$.

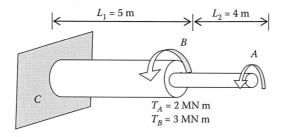

$$T_A = 2 \text{ MN m}$$
$$T_B = 3 \text{ MN m}$$

Solution

This is a statically determinate system subjected to only twist moments. The reaction torque at the built-in end can be determined from the equilibrium of moment. The method of section can be used to calculate the torques on the sections between A and B and B and C. The stresses on these sections can then be calculated by Equation 3.2. Because the shaft is under a discrete variation of torques and geometrical dimension, Equation 3.7 is used to calculate the angle of twist.

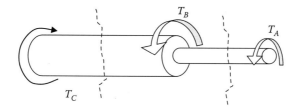

From the equilibrium of figure above:

$$T_C - T_A - T_B = 0$$
$$T_C = T_A + T_B = 5 \text{ MNm}$$

By taking sections between A and B and B and C (see figure above), the twist moment diagram is drawn in figure below. Following the right-hand rule, the twist moments on the sections along the axis are defined as positive.

Total angle of twist

$$\text{Since } G_{AB} = G_{BC} = G = \frac{E}{2(1 + v)} = \frac{200}{2(1 + 0.25)} = 80 \text{ GPa}$$

$$J_{AB} = \frac{\pi}{32}(0.5)^4 = \frac{\pi}{512} \text{ m}^4$$

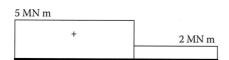

$$J_{BC} = \frac{\pi}{32}(1)^4 = \frac{\pi}{32} m^4$$

From Equation 3.6, the angle of twist at A is

$$\theta_A = \sum_{i=1,2} \frac{T_i L_i}{G_i J_i} = \frac{T_{AB} L_{AB}}{G_{AB} J_{AB}} + \frac{T_{BC} L_{BC}}{G_{BC} J_{BC}}$$

$$= \frac{2 \times 4}{80 \times 10^3 \pi/512} + \frac{5 \times 5}{80 \times 10^3 \pi/32} = 19.48 \times 10^{-3} \text{ rad}$$

The angle of twist at B is

$$\theta_B = \frac{T_{BC} L_{BC}}{G_{BC} J_{BC}} = \frac{5 \times 5}{80 \times 10^3 \pi/32} = 3.18 \times 10^{-3} \text{ rad}$$

Maximum shear stress

The maximum shear stresses within AB and BC are needed for the overall maximum. From Equation 3.2

$$\tau_{AB}^{max} = \frac{T_{AB} r_{max}}{J_{AB}} = \frac{2 \times (0.5/2)}{\pi/512} = 81.49 \text{ MPa}$$

$$\tau_{BC}^{max} = \frac{T_{BC} r_{max}}{J_{BC}} = \frac{5 \times (1/2)}{\pi/32} = 25.46 \text{ MPa}$$

Thus, the maximum shear stress occurs between A and B.

EXAMPLE 3.2

A motor drives a circular shaft through a set of gears at 630 rpm; 20 kW are delivered to a machine on the right and 60 kW on the left (figure below). If the allowable shear stress of the shaft is 37 MPa, determine the minimum diameter of the shaft.

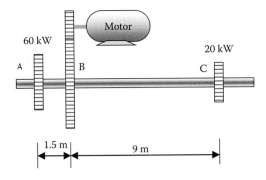

Solution

The torsion problem is statically determinate. The torques acting on the shaft have to be calculated first by Equation 3.9. The shaft is then analysed by Equation 3.2 to determine the maximum shear stress that depends on the diameter. A comparison of the maximum shear stress with the allowable stress of the material leads to the solution of the minimum diameter.

Since the total power transmitted to the shaft through gear B is 80 kW (60 + 20)

$$T_B = 9540 \frac{p}{\text{Rev}} = 9540 \times \frac{80}{630} = 1211.43 \,\text{N m}$$

The resistance torques at A and C are, respectively,

$$T_A = 9540 \frac{p}{\text{Rev}} = 9540 \times \frac{20}{630} = 302.86 \,\text{N m}$$

$$T_C = 9540 \frac{p}{\text{Rev}} = 9540 \times \frac{60}{630} = 908.57 \,\text{N m}$$

The twist moment diagram of the shaft is shown below.

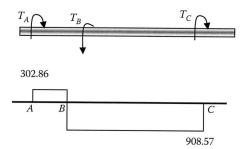

Because the shaft has a constant diameter, it is obvious that the maximum twist moment, then the maximum shear stress, occurs between B and C. By Equation 3.2

$$\tau_{max} = \frac{T_{max}D/2}{J}$$

$$= \frac{908.57(\text{N m}) \times D/2}{\pi D^4/32} \leq \tau_{allowable} = 37 \times 10^6 \,\text{N/m}^2$$

$$D \geq 5 \times 10^{-2} \,\text{m} = 50 \,\text{mm}$$

Thus, the diameter of the shaft must not be smaller than 50 mm.

EXAMPLE 3.3

A uniform steel pile of circular section, which has been driven to a depth of L in clay, carries an applied torque T at the top. This load is resisted entirely by the moment of friction m_f along the pile, which is linearly distributed along the depth as shown in figure below. If $L = 60$ m, the diameter of the pile $D = 100$ mm, $T = 4$ kN m, shear modulus $G = 80$ GPa and the allowable shear stress of material $\tau_{allowable} = 40$ MPa. Check the safety of the pile and determine the total angle of twist.

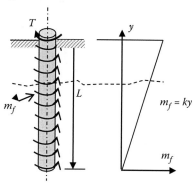

Solution

This is a statically determinate system. From the equilibrium requirement, the applied torque must be equal to the resultant torque due to the moment of friction, from which the constant k can be determined. The method of section can be used to find the torque on an arbitrary section and then draw the diagram of torque along the axis of the pile. The units used in the following calculation are m and N.

From the equilibrium of the pile, the resultant of m_f must equal the applied torque T. Thus, from figure above

$$T = \text{area of the triangle} = \frac{60 \times 60k}{2} = 4000 \, \text{N m}$$

$$k = \frac{4000 \times 2}{60 \times 60} = 2.22 \, \text{N m/m}$$

Take a cut at y (figure below) and consider equilibrium of the low segment.

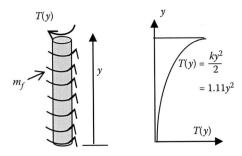

$$T(y) = \frac{ky \times y}{2} = 1.11y^2$$

The maximum torque occurs on the section at $y = 60$ m, that is, the section in level with the ground surface

$$\tau_{max} = \frac{T(L) \times D/2}{J} = \frac{4 \times 10^3 \times 50 \times 10^{-3}}{(\pi/32) \times 100^4 \times 10^{-12}}$$
$$= 20.37 \, \text{MPa} < 40 \, \text{MPa}$$

The design of the pile is satisfactory. Since the pile is subjected to the continuously distributed twist moment shown in figure above, from Equation 3.8

$$\theta_{total} = \int_0^L \frac{T(y)}{G(y)J(y)} \, dy = \int_0^L \frac{ky^2/2}{GJ} \, dy = \frac{k}{2GJ} \int_0^L y^2 dy$$

$$= \frac{kL^3}{6GJ}$$

$$= \frac{1.11 \times 60^3}{6 \times 80 \times 10^9 (\pi/32) \times 100^4 \times 10^{-12}}$$

$$= 0.051 \, \text{rad} = 2.92°$$

EXAMPLE 3.4

A bar of length 4 m shown in figure (a) below has a circular hollow cross section with an outside diameter D_{out} of 50 mm and is rigidly built in at each end. It carries torques of 0.9 and 1.5 kN m at the mid-span and three-quarter span sections, respectively, taken from the left-hand end. If both torques are applied in the same direction and the maximum shear stress in the bar is limited to 100 N/mm², calculate the maximum allowable internal diameter D_{in} of the bar.

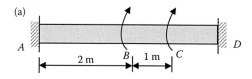

Solution

This is a statically indeterminate structure and hence the reaction torques on the cross sections at A and D cannot be uniquely determined by considering the equilibrium of the system. Thus a compatibility requirement must be considered, for example, by replacing the support at D with a twist moment T_D (figure below). Under the action of the three twist moments, the angle of twist of section D must be zero (fixed end condition). After the end moment at D is found, the system is equivalent to a statically determinate system. The procedure shown in previous examples can then be followed to determine the final solution.

Step 1: Static equilibrium equation
Assume that the reaction torques acting on the cross sections at A and D are, respectively, T_A and T_D. From the equilibrium of the bar (figure [b] below)

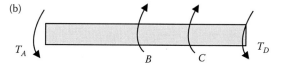

$$T_A + T_D - T_B - T_C = 0$$

$$T_A + T_D = T_B + T_C = 2.4 \text{ kN m}$$

Step 2: Compatibility equation
Now consider the angle of twist at D under the combined action of T_B, T_C and T_D as a superimposition of the angle of twist caused by the action of these torques individually, as shown in figures (c) to (f) below, respectively.

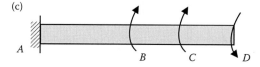

From figure (d) below and Equation 3.6

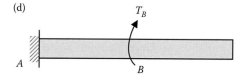

$$(\theta_D)_{(d)} = \theta_B = \frac{T_B L_{AB}}{GJ}$$

From figure (e) below and Equation 3.6

(e)

$$(\theta_D)_{(e)} = \theta_C = \frac{T_C L_{AC}}{GJ}$$

And from figure (f) below and Equation 3.6

(f)

$$(\theta_D)_{(f)} = \frac{T_D L_{AD}}{GJ}$$

The compatibility condition requires

$$(\theta_D)_{(d)} + (\theta_D)_{(e)} - (\theta_D)_{(f)} = 0$$

Step 3: Solution of the reaction torques at A and D

$$\frac{T_B L_{AB}}{GJ} + \frac{T_C L_{AC}}{GJ} - \frac{T_D L_{AD}}{GJ} = 0$$

$$0.9 \times 2 + 1.5 \times 3 - T_D \times 4 = 0$$

$$T_D = 1.57 \, \text{kN m}$$

From the equilibrium

$$T_A = 2.4 - T_D = 2.4 - 1.575 = 0.825 \, \text{kN m}$$

From the twist moment diagram shown in figure (g) below, the maximum twist moment occurs between C and D, where the design requirement must be met.

(g)

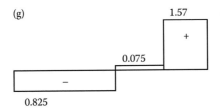

Step 4: Calculation of stress and the minimum diameter
Now from Equation 3.2

$$\tau = \frac{Tr}{J} = \frac{1.575 \times 10^6 \times D_o/2}{\pi(D_{out}^4 - D_{in}^4)/32}$$

Thus, for the maximum internal diameter

$$\frac{1.575 \times 10^6 \times D_{out}/2}{\pi(D_{out}^4 - D_{in}^4)/32} \leq 100$$
$$D_{in} = 38.7 \, \text{mm}$$

The internal diameter of the shaft must not be larger than 38.7 mm.

EXAMPLE 3.5

A tube of aluminium with a solid core of steel on the inside is shown in figure below. The member is subjected to a torque T. The shear moduli for aluminium and steel are, respectively, G_a and G_{steel}. What are the maximum shear stresses in the aluminium and steel?

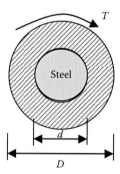

Solution

This is also a statically indeterminate problem. Due to the composite section, the shear stress distributions over the steel and aluminium areas have to be considered separately. The compatibility of deformation condition is that the angle of twist has the same value for both steel and aluminium.

Assume that the twist moments carried by the steel and aluminium are, respectively, T_{steel} and T_a. Thus, from equilibrium

$$T_a + T_{steel} = T$$

Since the steel and aluminium have the same angle of twist, from Equation 3.6

$$\frac{T_{steel}L}{G_{steel}J_{steel}} = \frac{T_aL}{G_aJ_a}$$

From the above two equations

$$T_a = \frac{G_aJ_a}{G_aJ_a + G_{steel}J_{steel}}T$$

$$T_{steel} = \frac{G_{steel}J_{steel}}{G_aJ_a + G_{steel}J_{steel}}T$$

Then, the maximum shear stresses in the steel and aluminium are, respectively,

$$\tau_a^{max} = \frac{T_a \times D/2}{J_a} = \frac{TG_aD}{2(G_aJ_a + G_{steel}J_{steel})}$$

$$\tau_{steel}^{max} = \frac{T_{steel} \times d/2}{J_{steel}} = \frac{TG_{steel}d}{2(G_aJ_a + G_{steel}J_{steel})}$$

EXAMPLE 3.6

A hollow steel shaft with an internal diameter of $d = 8$ in. and an outside diameter of $D = 12$ in. is to be replaced by a solid alloy shaft. If the maximum shear stress has the same value in both shafts, calculate the diameter of the latter and the ratio of the torsional rigidities GJ, where $G_{steel} = 2.4G_{alloy}$.

Solution

Because the maximum shear stress must be the same in both shafts under the same torque, from Equation 3.2, $[r_{max}/J]_{steel}$ must be equal to $[r_{max}/J]_{alloy}$.
Thus,

$$\frac{D_{steel}/2}{J_{steel}} = \frac{D_{alloy}/2}{J_{alloy}}$$

For the shafts,

$$J_{steel} = \frac{\pi}{32}(12^4 - 8^4) = 520\pi \text{ in.}^4$$

$$J_{alloy} = \frac{\pi}{32}D_{alloy}^4$$

We have

$$\frac{6}{520\pi} = \frac{D_{alloy}/2}{\frac{\pi}{32}D_{alloy}^4}$$

Hence, $D_{alloy}^3 = 1386 \text{ in.}^3$ and $D_{alloy} = 11.15$ in.
Ratio of torsional rigidity

$$\frac{G_{steel}J_{steel}}{G_{alloy}J_{alloy}} = \frac{G_{steel}}{G_{alloy}} \times \frac{D_{steel}}{D_{alloy}}$$

$$= 2.4 \times \frac{12}{11.15} = 2.58$$

That is, the torsional rigidity of the steel is 2.58 times that of the alloy shaft. This ratio means that, though replacing the steel shaft by the alloy one meets the strength requirement, the angle of twist of the alloy shaft will be 2.58 times greater than that of the steel.

EXAMPLE 3.7

When a bar of rectangular section is under torsion (figure below), determine the shear stresses at the corners of the cross section.

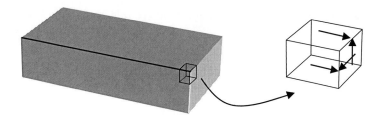

Solution

This question tests your conceptual understanding of shear stresses. They always appear in pairs on planes that are perpendicular to each other. They are equal in magnitude, but in an opposite sense.

Take an infinitesimal element around a corner point and assume that on the plane of the cross section there are two shear stresses perpendicular to either the horizontal or the vertical boundary. The two assumed shear stresses must be equal to the shear stresses acting on the top and the left-hand-side surfaces. Since the bar has a stress-free surface, the surface shear stresses are zero (Section 1.4.2). Consequently, the two assumed shear stresses at the corner of the cross section are also zero.

3.8 CONCEPTUAL QUESTIONS

1. Explain the terms 'shear stress' and 'shear strain' as applied to the cross section of a long shaft subjected to torsion.
2. Explain the terms 'torsional rigidity' and 'polar 2nd moment of area'.
3. Explain why Equations 3.2 to 3.8 can only apply for circular members.
4. Which of the following structures is under torsion?

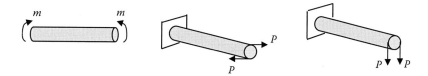

5. When a circular cross section is under pure torsion, why does the direction of the shear stress along the circular boundary always coincide with the tangent of the boundary?
6. When a twist moment is applied at a location along the axis of a bar, is it right to say that the twist moment on the cross section at the same location is equal to the external applied moment and why?
7. The twist moment diagrams for a shaft subjected to torsion are shown in figure below. Determine the magnitude, direction and location of the externally applied moments for each case.

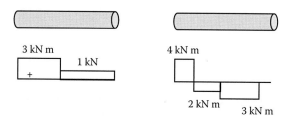

8. The tube sections below have the same outside radius and are subjected to the same twist moment.

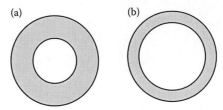

(a)   (b)

Which of the following statements are correct and why?
a. The maximum shear stresses of the two sections are the same.
b. The maximum shear stress of (a) is greater than that of (b).
c. The maximum shear stress of (b) is greater than that of (a).
d. The minimum shear stress of (b) is greater than that of (a).

9. Under torsion, the small square in figure below will change to

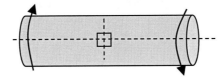

a. A larger square
b. A rectangle
c. A diamond
d. A parallelogram

10. Brittle materials normally fail at maximum tension. If the shaft shown in figure below is made of iron and subjected to pure torsion, which of the following statements is correct?

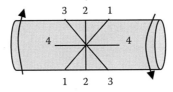

a. The shaft is likely to fail along section 1-1.
b. The shaft is likely to fail along section 2-2.
c. The shaft is likely to fail along section 3-3.
d. The shaft is likely to fail along section 4-4.

11. A solid circular section of diameter D can carry a maximum torque T. If the circular area of the cross section is doubled, what is the maximum torque that the new section can carry?
a. $\sqrt{2T}$
b. $2\sqrt{2T}$
c. $2T$
d. $4T$

12. A solid circular section has the same area as a hollow circular section. Which section has higher strength and stiffness, and why?

13. A solid circular shaft of steel and a solid circular shaft of aluminium are subjected to the same torque. If both shafts have the same angle of twist per unit length (θ/L),

which one of the following relationships will the maximum shear stresses in the shafts satisfy?

a. $\tau_{steel} < \tau_{aluminium}$.
b. $\tau_{steel} > \tau_{aluminium}$.
c. $\tau_{steel} = \tau_{aluminium}$.
d. All of the above are possible.

3.9 MINI TEST

PROBLEM 3.1

The twist moment diagrams are shown in figure below for a shaft subjected to torsion. Determine the magnitude, direction and location of the externally applied moments for each case.

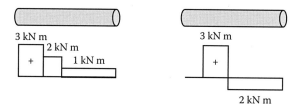

PROBLEM 3.2

Two geometrically identical shafts are loaded with the same twist moments. If the two shafts are made of different materials, which of the following statements is correct?

a. Both the maximum shear stresses and the angles of twist are the same.
b. Both the maximum shear stresses and the angles of twist are different.
c. The maximum shear stresses are the same, but the angles of twist are different.
d. The maximum shear stresses are different, but the angles of twist are the same.

PROBLEM 3.3

The hollow circular sections shown in figure below are subjected to torsion. Which of the following stress distributions is correct?

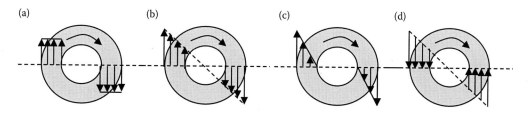

(a)   (b)   (c)   (d)

PROBLEM 3.4

Calculate the minimum diameters of two shafts transmitting 200 kW each without exceeding the allowable shear stress of 70 MPa. One of the shafts rotates at 20 rpm and the other at 20,000 rpm. What conclusion can you make from the relationship between the stresses and the speed of the shafts?

PROBLEM 3.5

Consider the stepped shaft shown in figure below. The shaft is fixed at both ends. Assuming that a, d, G and T are all given constants, determine the maximum shear stress and the angle of twist at B.

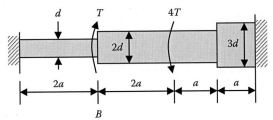

Chapter 4

Shear and bending moment

Figure 4.1a shows a car crossing a bridge. The purpose of the bridge deck is to transfer the weight of the car to the two supports. Figure 4.1b is an equivalent illustration of Figure 4.1a, showing the transversely applied load (the car), the structure (the deck), and the supports at the ends. In fact, in many engineering applications, structural members resist forces (loads) applied laterally or transversely to their axis. This type of member is termed a *beam*. The deformation of a beam is characterized as follows:

- The external load is applied transversely and causes the beam to flex as shown in Figure 4.1c.
- Bending moment and shear force develop on cross sections of the beam as shown in Figure 4.1d.
- Under the action of the bending moment and shear force, the deformation of the beam is dominated by transverse deflection and a rotation of cross section.
- Axial deformation of the beam is neglected.

4.1 DEFINITION OF BEAM

A beam is defined as a structural member designed primarily to support forces acting perpendicular to the axis of the member (Figure 4.2). The major difference between beams (Figure 4.2a) and axially loaded bars (Figure 4.2b) or shafts in torsion (Figure 4.2c) is in the direction of the applied loads.

4.2 SHEAR FORCE AND BENDING MOMENT

A shear force is an internal force that is parallel to the section it is acting on and that, for example, in Figure 4.1d, resists the vertical effect of the applied loads on the beam. The shear force is numerically equal to the algebraic sum of all the vertical forces acting on the free body taken from either sides of the section. Shear forces are measured in N, kN and so on.

A bending moment is an internal force that resists the effect of moments caused by external loads, including support reactions. Bending moments are measured in N m, kN m and the like.

4.3 BEAM SUPPORTS

A beam may be supported differently. Table 4.1 shows the most common types of supports that are frequently used to model practical structural supports in design.

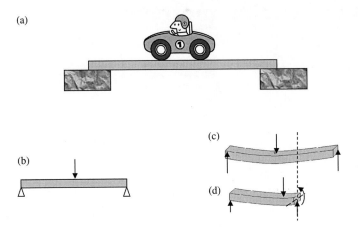

Figure 4.1 Illustration of beam bending.

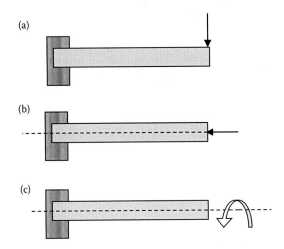

Figure 4.2 Types of deformation.

4.4 SIGN CONVENTION

The following sign convention is adopted for bending analysis of beams in most textbooks. The convention is again not in relation to a particular coordinate direction.

4.4.1 Definition of positive shear

A downward shear force acting on the cross section of the left-hand-side free body of a cut, or an upward shear force acting on the cross section of the right-hand-side free body of the cut, is defined as a positive shear force (Figure 4.3a). Positive shear forces are shown in Figure 4.3b for a segment isolated from a beam by two sections (cuts).

The shear forces shown in Figure 4.3b tend to *push up* the left-hand side of the beam segment. The positive shear forces can also be described as *left up* shear forces, causing rotation of the segment in a clock-wise manner.

Table 4.1 Beam supports and reactions

Type of support		Simple illustration	Reaction forces and displacements at the support	
Roller		or	1. Shear force	1. Axial displacement (usually ignored[a]) 2. Rotation
Pinned			1. Shear force 2. Axial force (usually ignored[a])	1. Rotation
Fixed			1. Shear force 2. Axial force (usually ignored[a]) 3. Bending moment	None
Free			None	1. Axial force (usually ignored[a]) 2. Vertical displacement 3. Rotation

[a] Axial force/displacement is significantly smaller than other types of forces/displacements in bending.

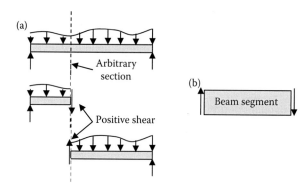

Figure 4.3 Sign convention of shear force.

4.4.2 Definition of positive bending moment

A positive bending moment (Figure 4.4a) produces compression in the upper part and tension in the lower part of a beam's cross section. The deformed beam takes a shape that can *retain water.*

From Figure 4.4b, a positive bending moment can also be described as *sagging moment* since the moment induces a sagging deflection.

4.5 RELATIONSHIPS BETWEEN BENDING MOMENT, SHEAR FORCE AND APPLIED LOAD

On an arbitrary cross section of a beam shown in Figure 4.5, the interrelation of the bending moment $M(x)$, shear force $V(x)$ and the intensity of the distributed load $q(x)$ always conform to the following relationships:

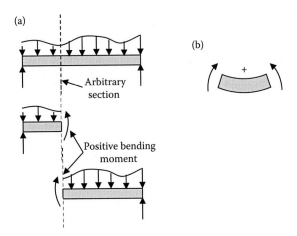

Figure 4.4 Sign convention of bending moment.

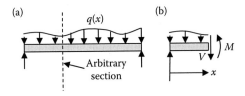

Figure 4.5 Internal forces of loaded beam.

1. The change rate of shear force along a beam is equal to the distributed load

$$\frac{dV(x)}{dx} = -q(x) \qquad (4.1)$$

2. The change rate of bending moment along a beam is equal to the shear force

$$\frac{dM(x)}{dx} = V(x) \qquad (4.2)$$

3. The combination of Equations 4.1 and 4.2 yields

$$\frac{d^2M(x)}{dx^2} = -q(x) \qquad (4.3)$$

The above relationships are applicable at all cross sections of a beam except where there is a concentrated force or moment.

4.6 SHEAR FORCE AND BENDING MOMENT DIAGRAMS

The shear force and bending moment diagrams show, respectively, the internal shear force and bending moment distribution on the cross section along the length of a beam.

On the basis of the interrelations shown in Section 4.5, Table 4.2 shows some common features exhibited in the shear force and bending moment diagrams of beams.

Table 4.2 Common features of shear and bending moment diagrams

	Case (a)	Case (b)	Case (c)	Case (d)
Load acting on a segment of a beam	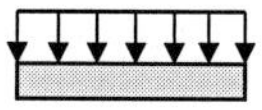	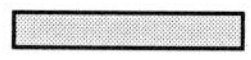	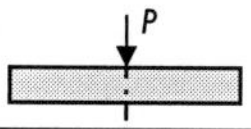 P	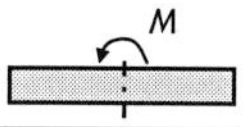 M
Shear force V $$\frac{dV}{dx}=-q$$		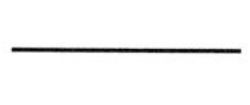	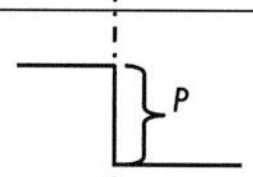	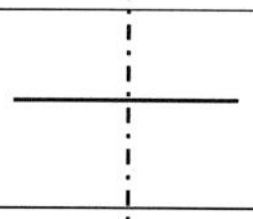
Bending moment M $$\frac{d^2M}{dx^2}=-q$$		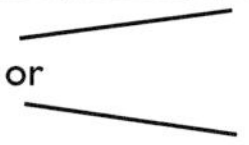 or	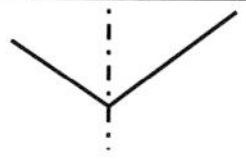	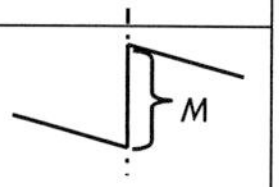 M
Observation	1. Linearly decreasing shear force from left to right (may change from positive to negative). 2. Bending moment diagram is similar to the shape of a string under a uniformly distributed load, i.e., a parabola.	1. Constant positive or negative shear force. 2. Linearly increasing or decreasing bending moment, depending on the loads acting on other parts of the beam.	1. An abrupt drop of shear force by P when passing through the section where the point load is applied. 2. Bending moment diagram showing the shape of a string subjected to a point load.	1. Continuous distribution of shear force across the section where the external moment is applied. 2. An abrupt increase of bending moment by M when passing through the section.

4.7 KEY POINTS REVIEW

- A beam is long in the axial direction compared to its other two dimensions.
- A beam is to support external load applied perpendicular to the axis.
- The two major internal forces are shear force and bending moment.
- The change in shear force is equal to applied distributed load.
- The change in bending moment is equal to shear force.
- A shear force diagram shows how shear force is distributed along the axis of a beam.
- A bending moment diagram shows how bending moment is distributed along the axis of a beam.
- Continuous distribution of applied loads results in continuous variations of both shear force and bending moment along a beam.
- On unloaded segments of a beam the shear force is constant.
- A uniformly distributed load causes a linear variation in shear force, and a parabolic variation in bending moment.
- A concentrated force/moment results in sudden drop or jump of shear force/bending moment at the location where the force/moment is applied.
- A point force causes a *kink* in a bending moment diagram.

- A concentrated moment has no effect on a shear force diagram.
- There is no shear force at a free end of a beam.
- There is no bending moment at a free or simply supported end of a beam.
- An internal hinge can transmit a shear force.
- An internal hinge cannot transmit a bending moment.
- Bending moment reaches either maximum or minimum at a point of zero shear force.

4.8 RECOMMENDED PROCEDURE OF SOLUTION

- Replace all supports of a beam by their associated reactions (see Table 4.1).
- Apply the static equilibrium equations to determine the reactions.
- Identify *critical sections* that characterize changes of pattern of the internal force diagrams. The critical sections include the locations where
   - A concentrated force or moment is applied
   - A beam is supported
   - A distributed load starts or ends
- Apply the method of section by taking cuts between each of the critical sections.
- Take either left or right part of a cut as a free body and repeat this for all the cuts taken.
- Add unknown shear force and bending moment on all the cuts, assuming that they are all positive by following the sign convention.
- Consider static equilibrium of the free body taken by each of the cuts and calculate the shear forces and bending moments on the cross sections (cuts).
- Use Table 4.2 to draw the diagrams between the critical sections. If a moment distribution is parabolic between two critical sections and accurate distribution is required, at least an additional cut must be taken between the two sections.

4.9 EXAMPLES

EXAMPLE 4.1
Draw the shear force and bending moment diagrams of the simply supported beam shown in figure below.

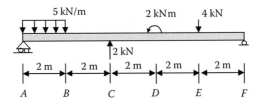

Solution
This is a determinate problem that can be solved by following the general procedure described in Section 4.8.

Step 1: Replace supports by reactions

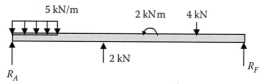

Since there is no external load applied horizontally, the horizontal support reaction at A vanishes. Thus, the pin at A is replaced by a single reaction R_A in the vertical direction.

Step 2: Solution of the reactions
Taking anticlockwise moments about A

$$\Sigma M_A = 10\,m \times R_F + 2\,kN \times 4\,m + 2\,kNm - 4\,kN \times 8\,m - 5\,kN/m \times 2\,m \times 1\,m = 0$$
$$R_F = 3.2\,kN\,(\uparrow)$$

Resolving vertically $(\uparrow)$

$$R_A + R_F + 2\,kN - 4\,kN - 5\,kN/m \times 2\,m = 0$$
$$R_A = 8.8\,kN\,(\uparrow)$$

Step 3: Identify critical sections
Sections at A, B, C, D, E and F are all critical sections, where either application of concentrated forces or change of load pattern occurs.

Step 4: Calculation of shear forces and bending moments on the critical sections
1. Section at A
At section A (the left-end section), R_A acts as a positive shear force (left up), while there is no bending moment. Thus,

$$V_A = R_A = 8.8\,kN$$
$$M_A = 0$$

2. Section at B

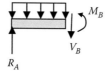

Resolving vertically

$$R_A - 5\,kN/m \times 2\,m - V_B = 0$$
$$V_B = -1.2\,kN\,(\uparrow)$$

The negative sign denotes that the actual direction of V_B is opposite to the assumed direction.
Taking anticlockwise moment about B

$$M_B + 5\,kN/m \times 2\,m \times 1\,m - R_A \times 2\,m = 0$$
$$M_B = 7.6\,kNm\,(\circlearrowleft)$$

In order to draw the parabolic distribution of the bending moment between A and B, an additional section between A and B must be considered. Let us take the middle span of AB.

Taking anticlockwise moment about the cut

$$M_{AB} + 5\,\text{kN/m} \times 1\,\text{m} \times 0.5\,\text{m} - R_A \times 1\,\text{m} = 0$$
$$M_{AB} = 6.3\,\text{kN}\,\text{m}\ (\circlearrowleft)$$

3. Section at C

We can consider the section on the immediate left or right of the concentrated load applied at C, that is, with or without including the concentrated load in the free body diagram (FBD). We take the section on the left of the load at C in the following calculations.

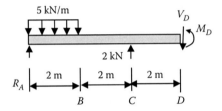

Resolving vertically

$$R_A - 5\,\text{kN/m} \times 2\,\text{m} - V_C = 0$$
$$V_C = -1.2\,\text{kN}\ (\uparrow)$$

Taking anticlockwise moment about C

$$M_C - 5\,\text{kN/m}\ 2\,\text{m} \times 3\,\text{m} - R_A \times 4\,\text{m} = 0$$
$$M_C = 5.2\,\text{kN}\,\text{m}\ (\circlearrowleft)$$

4. Section at D

Once again, we can consider the section on either the immediate left or the immediate right of the concentrated moment applied at D, that is, with or without including the moment in the FBD. We take the section on the left of the moment at D in the following calculations.

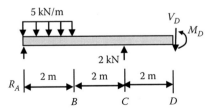

Resolving vertically

$$R_A + 2\,\text{kN} - 5\,\text{kN/m} \times 2\,\text{m} - V_D = 0$$
$$V_D = 0.8\,\text{kN}\ (\downarrow)$$

Taking anticlockwise moment about D

$$M_D + 5\,\text{kN/m} \times 2\,\text{m} \times 5\text{m} - 2\,\text{kN} \times 2\,\text{m} - R_A \times 6\,\text{m} = 0$$
$$M_D = 6.8\,\text{kN}\,\text{m}\ (\circlearrowleft)$$

5. Section at E

We take the section on the left-hand side of the concentrated load applied at E.

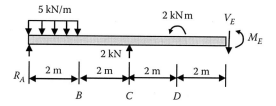

Resolving vertically

$$R_A + 2\,\text{kN} - 5\,\text{kN/m} \times 2\,\text{m} - V_E = 0$$
$$V_E = 0.8\,\text{kN}\ (\downarrow)$$

Taking anticlockwise moment about E

$$M_E + 2\,\text{kN}\,\text{m} + 5\,\text{kN/m} \times 2\,\text{m} \times 7\,\text{m} - 2\,\text{kN} \times 4\,\text{m} - R_A \times 8\,\text{m} = 0$$
$$M_E = 6.4\,\text{kN}\,\text{m}\ (\circlearrowleft)$$

6. Section at F

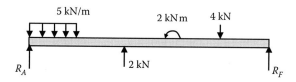

The only reaction at F is the vertically upward force R_F, that is, the shear force on the section. Thus,

$$V_E = R_F = -3.2\,\text{kN}$$
$$M_F = 0$$

The shear force is negative by following the sign convention established in Section 4.4.1.

Step 5: Shear force and bending moment diagrams

From Table 4.2, the shear force diagram between A and B is a straight line valued 8.8 kN at A and −1.2 kN at B, while the bending moment diagram between A and B is a parabola (case (a) in Table 4.2). We have known the values of the bending moment at three locations from A to B, that is, at A, B and the middle point of AB. The accurate shape of the parabola can then be drawn.

Between B and C (case (b) in Table 4.2), the shear force is a constant that equals the shear force at B (there is no concentrated force applied here), that is, $V_B = -1.2$ kN. The bending moment diagram is a straight line valued 7.6 kN m at B and 5.2 kN m at C (there is no concentrated moment applied at B). Since there are no concentrated force and moment applied at B, both the shear force and the bending moment show no abrupt changes across B.

Between C and D (case (b) in Table 4.2), the shear force is a constant valued 0.8 kN (from $V_D = 0.8$ kN), while the bending moment varies linearly with a value of 6.8 kN m at D. Obviously, due to the concentrated force 2 kN applied at C, the shear force shows an abrupt jump in the direction of the applied force when crossing C from the left-hand side of the force (case (c) in Table 4.2). Since there is no concentrated moment applied at C, the bending moment diagram is continuous across section C.

Between D and E (case (b) in Table 4.2), the shear force and bending moment are, respectively, constant and linear. Since there is no concentrated load applied at D, the constant shear between D and E is equal to the shear force between C and D, that is, 0.8 kN. The bending moment at D has an abrupt jump of 2 kN m (case (d) in Table 4.2) due to the concentrated moment applied at D. The moment then varies linearly and is 6.4 kN m at E ($M_E = 6.4$ kN m).

Between E and F (case (b) in Table 4.2), the shear force is constant with an abrupt drop from 0.8 kN to −3.2 kN at E due to the applied downward load of 4 kN (case (c) in Table 4.2). The bending moment diagram is linear between E and F. Since there is no concentrated moment applied at E, the bending moment continues across the section. The bending moment vanishes as $M_F = 0$ at the section supported by the roller pin.

On the basis of the calculations in Step 4 and the analysis in Step 5, the shear force and bending moment diagrams of the beam are drawn below.

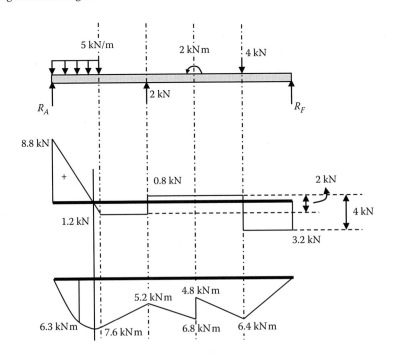

The following are observations from the above diagrams:

1. When shear force is a constant over a span of a beam, the bending moment over the same span is a sloped straight line (see Equation 4.2).
2. When shear force is a sloped straight line over a span of a beam, the bending moment over the same span is a parabola (see Equation 4.2).
3. At a point where shear force diagram passes through zero, the bending moment is either maximum or minimum.

EXAMPLE 4.2

Draw the shear force and bending moment diagrams of the beam shown in figure below. The beam has a pin joint at B.

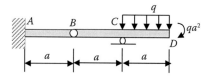

Solution

This is a statically determinate beam. The pin at B joints AB and BD. Thus, the bending moment is zero at B. From Table 4.2, we know that the shear force is constant and the bending moment varies linearly from A to C since there is no external load applied within this range. The shear force and the bending moment are, respectively, linear and parabolic between C and D. An abrupt change of shear force occurs at C due to the concentrated support reaction.

Step 1: Replace supports with reactions

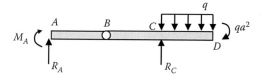

The horizontal reaction at A vanishes in this case since there is no external load applied horizontally.

Step 2: Calculate support reactions
Consider BD and take anticlockwise moment about B, noting that at B the bending moment is zero due to the hinge.

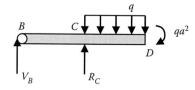

$$-qa^2 + R_Ca - qa \times \left(a + \frac{a}{2}\right) = 0$$

$$R_C = \frac{5qa^2}{2}(\uparrow)$$

Considering AD and resolving vertically

$$R_A + R_C - qa = 0$$

$$R_A = qa - R_C = -\frac{3qa}{2}(\downarrow)$$

Consider AB and take anticlockwise moment about B

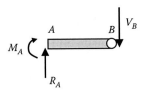

$$-R_A a - M_A = 0$$

$$M_A = -R_A a = -\left(-\frac{3qa}{2}\right)a = \frac{3qa^2}{2}(\circlearrowleft)$$

Step 3: Identify critical sections
Sections at A, C and D are the critical sections.

Step 4: Calculate shear forces and bending moments on the critical sections
1. Section at A
From the calculation in Step 2

$$V_A = R_A = -\frac{3qa}{2}$$

$$M_A = \frac{3qa^2}{2}$$

2. Section at C
Take the section on the left of the roller pin at C.

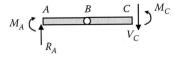

Resolving vertically

$$R_A - V_C = 0$$

$$V_C = R_A = -\frac{3qa}{2}(\uparrow)$$

Taking anticlockwise moment about C

$$M_C - M_A - R_A \times 2a = 0$$

$$M_C = M_A + 2aR_A = \frac{3qa^2}{2} + 2a \times \left(-\frac{3qa}{2}\right) = -\frac{3qa^2}{2} \,(\curvearrowleft)$$

3. Section at D

The section at D is a free end subjected to a concentrated moment. Thus,

$$V_D = 0$$
$$M_D = -qa^2$$

Following the sign convention, the moment at D is defined as negative.

Step 5: Draw shear force and bending moment diagrams

On the basis of the calculations in Step 4 and the discussion before Step 1, the shear force and bending moment diagrams are drawn below.

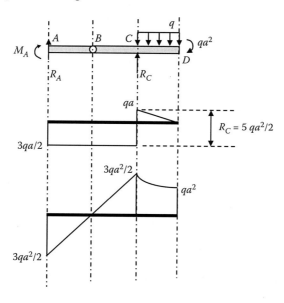

The following are observations from the above shear force and bending moment diagrams:

1. The bending moment is zero when passing through a pin joint.
2. A pin joint does not alternate the variation of shear force.
3. An abrupt change of shear force occurs on the section where an intermediate support is applied.

EXAMPLE 4.3

The beam shown in figure below is pinned to the wall at A. A vertical bracket BD is rigidly fixed to the beam at B, and a tie ED is pinned to the wall at E and to the bracket at D. The beam AC is subjected to a uniformly distributed load of 2 tons/ft and a concentrated load of 8 tons at C. Draw the shear force and bending moment diagrams for the beam.

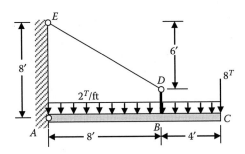

Solution

This is a statically determinate structure. The axial force in the tie can be resolved into vertical and horizontal components acting at D. The two force components generate a vertical force and a concentrated moment applied at the point B of the beam.

Step 1: Calculate force and moment at B and reaction forces

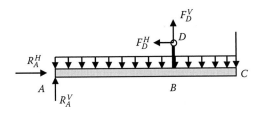

Resolving vertically

$$R_A^V + F_D^V - 8 \text{ tons} - 2 \text{ tons/ft} \times 12 \text{ ft} = 0$$

Taking anticlockwise moment about A

$$F_D^V \times 8 \text{ft} + F_D^H \times 2 \text{ft} - 8 \text{ tons} \times 12 \text{ ft} - 2 \text{ tons/ft} \times 12 \text{ ft} \times 6 \text{ ft} = 0$$

Due to the fact that the resultant of F_D^H and F_D^V must be in line with DE

$$\frac{F_D^V}{F_D^H} = \frac{6\,\text{ft}}{8\,\text{ft}} = \frac{3}{4}$$

or

$$F_D^V = \frac{3}{4} F_D^H$$

Considering the above three equations, we have

$$F_D^H = 30 \text{ tons } (\leftarrow)$$
$$F_D^V = 22.5 \text{ tons } (\uparrow)$$
$$R_A^V = 9.5 \text{ tons } (\uparrow)$$

The horizontal reaction at A, R_A^H, does not induce any shear or bending of the beam and is not included in the following calculation. The horizontal force, F_D^H, at D induces a concentrated moment applied at B. The moment caused by F_D^H is

$$F_D^H \times 2 \text{ ft} = 30^T \times 2 \text{ ft} = 60 \text{ tons} \cdot \text{ft}$$

Hence, the beam is loaded as

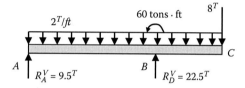

Step 2: Calculate shear forces and bending moments on critical cross sections. The cross sections at A, B and C are the critical sections.

1. Section at A

$$M_A = 0$$
$$V_A = R_A^V = 9.5 \text{ tons}$$

2. Section at B
   Taking cut at the left-hand side of B

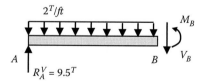

Resolving vertically

$$R_A^V - 2 \text{ tons/ft} \times 8 \text{ ft} - V_B = 0$$
$$V_B = R_A^V - 2 \text{ ton/ft} \times 8 \text{ ft} = -6.5 \text{ tons } (\uparrow)$$

Taking anticlockwise moment about B

$$-R_A^V \times 8 \text{ ft} + 2 \text{ tons/ft} \times 8 \text{ ft} + M_B = 0$$
$$M_B = R_A^V \times 8 \text{ ft} - 2 \text{ tons/ft} \times 4 \text{ ft} = 12 \text{ tons} \cdot \text{ft}$$

3. Section at C

$$M_C = 0 \quad \text{and}$$
$$V_C = 8 \text{ tons}$$

Step 3: Draw the shear force and bending moment diagrams

Due to the distributed load, the shear force diagrams between A and B and B and C are both sloping lines, while the bending moment diagrams are parabolas. At B, due to the concentrated force and bending moment, abrupt changes occur in both diagrams.

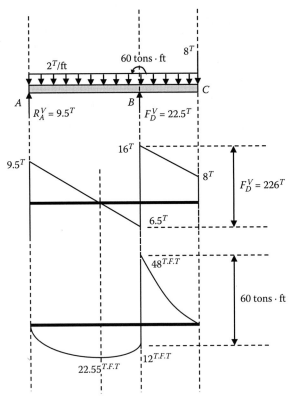

EXAMPLE 4.4

The simply supported beam shown in figure below is loaded with the triangularly distributed pressure. Draw the shear force and bending moment diagrams of the beam.

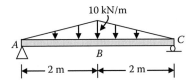

Solution

This is statically determinate beam subjected to linearly distributed load. From Table 4.2, a uniformly distributed load produces a linear shear force and a parabolic bending moment diagram. From the relationship shown in Equations 4.1 and 4.2, the shear force and bending moment diagrams of this case are, respectively, parabolic and cubic.

Step 1: Calculate support reactions

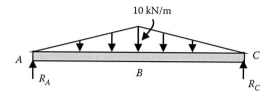

Resolving vertically

$$R_A + R_C = 2 \times \left(\frac{1}{2} \times 2\,\text{m} \times 10\,\text{kN/m} \right) = 20\,\text{kN}$$

Taking anticlockwise moment about A

$$R_C \times 4 - 20\,\text{kN} \times 2\,\text{m} = 0$$
$$R_C = 10\,\text{kN} \,(\uparrow)$$

Thus,

$$R_A = 10\,\text{kN} \,(\uparrow)$$

Step 2: Calculate shear forces and bending moments on critical sections
Sections at A, B (change in distribution pattern) and C are the critical sections.
  1. Section at A (pin)

$$V_A = R_A = 10\,\text{kN} \,(\uparrow)$$
$$M_A = 0$$

  2. Section at the immediate left of B

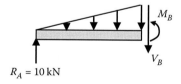

Resolving vertically

$$R_A - V_B - \frac{1}{2} \times 2\,\text{m} \times 10\,\text{kN/m} = 0$$
$$V_B = \frac{1}{2} \times 2\,\text{m} \times 10\,\text{kN/m} - 10\,\text{kN} = 0$$

Taking anticlockwise moment about B

$$M_B - R_A \times 2\,\text{m} + \frac{1}{2} \times 2\,\text{m} \times 10\,\text{kN/m} \times \frac{2}{3} \times 2\,\text{m} = 0$$

$$M_B = 13.33\,\text{kN m} \ (\curvearrowright)$$

3. At C (roller)

$$V_C = -R_C = -10\,\text{kN} \ (\uparrow)$$
$$M_C = 0$$

Step 3: Draw the shear force and bending moment diagrams
   The shear force and bending moment diagrams between the critical sections can be sketched as parabolas and cubes, respectively. If more accurate curves are required, shear forces and bending moments on additional sections between each pair of the consecutive critical sections are needed. For this purpose, it is appropriate to establish the general equations of shear force and bending moment between these critical sections.
   Between A and B: Take an arbitrary section between A and B, assume the distance from the section to A is x and consider the equilibrium of the beam segment shown in the figure below.

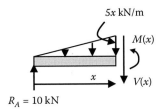

Resolving vertically

$$R_A - \frac{1}{2} \times x \times 5x - V(x) = 0$$

$$V(x) = 10 - \frac{5}{2}x^2 (\text{kN}) \ (\downarrow) \quad (0 \le x \le 2\,\text{m})$$

Taking anticlockwise moment about the section

$$M(x) - R_A \times x + \frac{1}{2} \times x \times 5x \times \frac{1}{3} \times x = 0$$

$$M(x) = 10x - \frac{5}{6}x^3 \quad (0 \le x \le 2\,\text{m})$$

Between B and C: Take an arbitrary section between B and C. Assume the distance from the section to C is x and consider the equilibrium of the beam segment shown in the following figure.

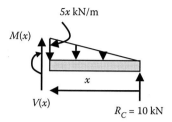

Resolving vertically

$$V(x) + R_C - \frac{1}{2} \times x \times 5x = 0$$

$$V(x) = -10 + \frac{5}{2}x^2 (\text{kN}) \quad (0 \le x \le 2\,\text{m})$$

Taking anticlockwise moment about the arbitrary section

$$M(x) - R_C \times x + \frac{1}{2} \times x \times 5x \times \frac{1}{3} \times x = 0$$

$$M(x) = 10x - \frac{5}{6}x^3 \quad (0 \le x \le 2\,\text{m})$$

The shear force and bending moment on an arbitrary section can be calculated by introducing the x coordinate into the general equation derived above. For example,
At $x = 1$ m from A between A and B

$$V(x) = 10 - \frac{5}{2}x^2 = 10 - \frac{5}{2} \times 1^2 = 7.5\,(\text{kN})\,(\downarrow)$$

$$M(x) = 10x - \frac{5}{6}x^3 = 10 \times 1 - \frac{5}{6} \times 1^3 = 9.17\,\text{kN/m}\,(\curvearrowright)$$

At $x = 1.2$ m from C between B and C

$$V(x) = -10 + \frac{5}{2}x^2 = -10 + \frac{5}{2} \times 1.2^2 = -6.4\,(\text{kN})\,(\downarrow)$$

$$M(x) = 10x - \frac{5}{6}x^3 = 10 \times 1.2 - \frac{5}{6} \times 1.2^3 = 10.56\,\text{kN/m}\,(\curvearrowright)$$

Hence, the shear force and bending moment diagrams are sketched as follows:

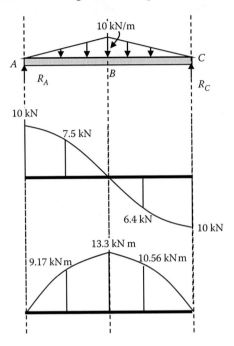

The following are observations from the above shear force and bending moment diagrams:

1. If a beam and the applied loads are both symmetric about the cross section at the mid-span, the shear stress is zero on the mid-span section and distributed antisymmetrically along the beam.
2. If a beam and the applied loads are both symmetric about the cross section at the mid-span, the bending moment diagram is symmetric about the mid-span section.
3. The distribution of shear force along the axis of a beam is one order higher than the order of the distributed load.
4. The distribution of bending moment along the axis of a beam is two order higher than the order of the distributed load.

EXAMPLE 4.5

The column–beam system shown in figure below is subjected to a horizontal pressure q. Draw the shear force and bending moment diagrams.

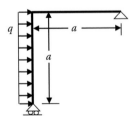

Solution

This is a statically determinate frame system. The internal forces can be computed by following the general procedure described in Section 4.8 for single beams.

Step 1: Replace supports by reactions

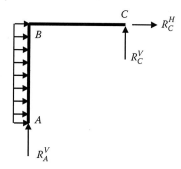

Due to the roller pin at A, the horizontal reaction at the support is zero.

Step 2: Derive solution of the reactions
Taking anticlockwise moment about C

$$qa \times \frac{a}{2} - R_A^V \times a = 0$$

$$R_A^V = \frac{qa}{2}$$

Resolving vertically

$$R_A^V + R_C^V = 0$$

$$R_C^V = -R_A^V = -\frac{qa}{2}$$

Step 3: Identify critical sections
Sections at A, B and C are critical sections. In addition to the critical sections specified in Section 4.8, the section at which there is a sudden change in member orientation is also taken as a critical section.

Step 4: Calculate shear forces and bending moments on the critical sections

1. Section at A
   The section is supported by a roller pin. Since $R_A^H = 0$, there is no shear force. The bending moment at the support is also zero. Thus,

   $$V_A = 0$$
   $$M_A = 0$$

2. Section at B

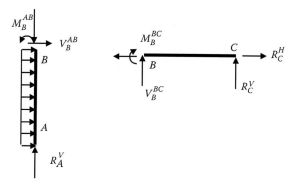

For the column, resolving horizontally

$$V_B^{AB} + qa = 0$$
$$V_B^{AB} = -qa$$

Taking anticlockwise moment about B

$$M_B^{AB} + qa\frac{a}{2} = 0$$
$$M_B^{AB} = -\frac{qa^2}{2}$$

For the beam, resolving vertically

$$V_B^{BC} + R_C^V = 0$$
$$V_B^{BC} = -R_C^V = \frac{qa}{2}$$

Taking anticlockwise moment about B

$$a \times R_C^V - M_B^{BC} = 0$$
$$M_B^{BC} = -aR_C^V = -\frac{qa^2}{2}$$

3. Section at C
   The pin at C prevents the section from any vertical displacement. The vertical support reaction is the shear force. The bending moment is zero:

$$V_C = R_C^V = -\frac{qa}{2}$$
$$M_C = 0$$

Step 5: Draw the shear force and bending moment diagrams

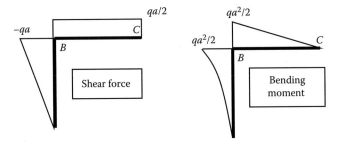

In the above diagrams, the sign of shear force is shown, while the bending moment diagram is drawn along the tension side of each member.

1. The internal force diagram of a frame can be drawn by following exactly the same procedure followed for a single beam.
2. At a rigid joint of two members, the bending moment is constant across the joint if there is no external moment applied at the location.
3. At a rigid joint of two members, the shear force is not necessarily constant across the joint even if there is no concentrated load applied at the location.

4.10 CONCEPTUAL QUESTIONS

1. What are the conditions for a beam to be in equilibrium?
2. What is a 'cantilever' and how is its equilibrium maintained?
3. What is a 'simply supported beam' and how is its equilibrium maintained?
4. Explain the terms 'shear force' and 'bending moment'.
5. Explain how the shear force and bending moment can be found on a section of a beam.
6. What are shear force and bending moment diagrams?
7. What are the relationships between shear force, bending moment and distributed load, and how can the relationships be used in plotting shear forces and bending moment diagrams?
8. A cantilever is loaded with a triangularly distributed pressure. Which of the following statements is correct?
   a. The shear force diagram is a horizontal line and the bending moment diagram is a sloping line.
   b. The shear force diagram is a sloping line and the bending moment diagram is a parabola.
   c. The shear force diagram is parabolic and the bending moment diagram is cubic.
   d. Both the shear force and the bending moment diagrams are parabolic.
9. A span of a beam carries only concentrated point loads, the shear force diagram is a series of _____ and the bending moment diagram is a series of _____.
10. The shear force diagram of a simply supported beam is shown in figure below. Which of the following observations are not correct?

   a. There is a concentrated force applied on the beam.
   b. There is no concentrated moment applied on the beam.
   c. There is a uniformly distributed load applied on the beam.
   d. There is a linearly distributed load applied on the beam.
11. The bending moment diagram of a simply supported beam is shown in figure below. Identify the types of loads applied on the beam.

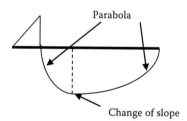

Parabola

Change of slope

12. Consider a cantilever subjected to concentrated point loads. The maximum bending moment must occur at the fixed end (Yes/No).
13. At a point between the two ends of a beam where the shear force diagram passes through zero, the bending moment is either a maximum or a minimum (Yes/No).
14. Over any part of a beam where the shear force is zero, the bending moment has a constant value (Yes/No).

15. The cantilever is loaded as shown in figure below. Which of the following four bending moment diagrams is correct?

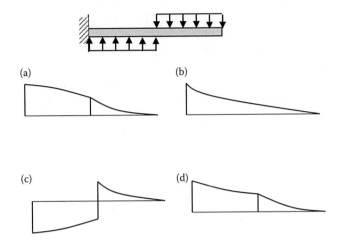

(a)

(b)

(c)

(d)

4.11 MINI TEST

PROBLEM 4.1

The shear force diagram over a segment of a beam is shown in figure below. Which of the following statements is not correct?

Parabola

1. A concentrated point load is applied.
2. A uniformly distributed load is applied.
3. A concentrated moment may be applied.
4. A triangularly distributed load is applied.
5. No concentrated moment is applied.

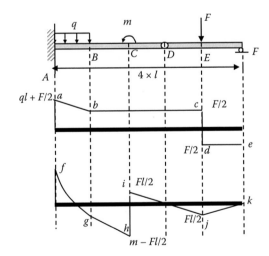

PROBLEM 4.2

From the loaded beam and its shear force and bending moment diagrams shown in figure below, complete the following statements:

1. The difference between the shear forces at c and d is _____.
2. The slope of line a–b is _____.
3. The difference between the bending moments at h and i is _____.
4. The slopes of line i–j and line j–k are, respectively, _____ and _____.

PROBLEM 4.3

A simply supported beam is subjected to a total of load P that is distributed differently along the beam as shown in figure below. Calculate the maximum shear forces and maximum bending moments for each case and compare the results. What conclusion can you draw from the comparisons?

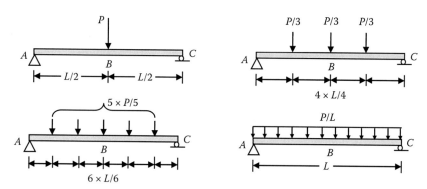

PROBLEM 4.4

For the beam loaded as shown in figure below, draw the shear force and bending moment diagrams.

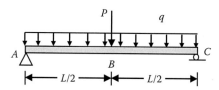

PROBLEM 4.5

Plot the shear force and bending moment diagrams of the beam loaded as shown in figure below. Sketch the curve of deflection of the beam.

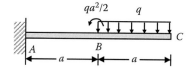

Chapter 5

Bending stresses in symmetric beams

In a beam subjected to transverse loads applied within the plane of symmetry (Figure 5.1a), only a bending moment and a shear force develop on the cross section (Figure 5.1b). The bending moment and shear force are, respectively, the resultants of the normal stresses (Figure 5.1c) and the shear stresses (Figure 5.1d) on the cross section. The deformation of the beam is characterised by the following:

- The longitudinal fibres on the top side of the beam contract (are shortened).
- The longitudinal fibres on the bottom side of the beam extend (are elongated).
- Between the top and the bottom sides, there is a parallel surface within which fibres are neither contracted nor extended.
- Due to the uneven elongation of the fibres, the beam exhibits lateral (transverse) deformation that is termed as *deflection*.

Two important beam deformation terminologies (Figure 5.2) are introduced on the basis of the above conceptual analysis of beam deformation.

- The surface formed by the fibres that are neither contracted nor extended is called *neutral surface*. This surface lies inside the beam between the top and the bottom surfaces. The beam fibres under the neutral surface are stretched and are in tension, while the beam fibres above the neutral surface are compressed and thus in compression.
- The intersection of the neutral surface and a cross section is called *neutral axis*. This intersection is a line within the cross section and passes through the centroid of the section.

Apparently, on the cross section, the area under the neutral axis is in tension, while the area above the neutral axis is in compression.

5.1 ELASTIC NORMAL STRESSES IN BEAMS

The elastic normal stresses on a cross section of a beam subjected to bending are calculated on the basis of the following basic assumptions:

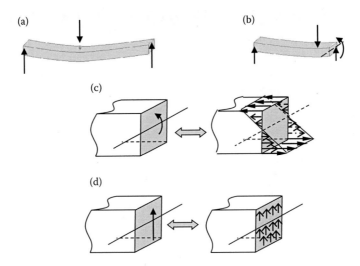

Figure 5.1 Stress distribution on cross section subjected to bending.

- The deformation of the cross section is within elastic range.
- A cross-sectional plane, taken normal to the beam's axis, remains plane throughout the bending deformation.
- The longitudinal strains of the fibres vary linearly across the depth, proportional to the distances from the fibres to the neutral axis.
- Hooke's law is applicable to the individual fibres, that is, stress is proportional to strain.

The general expression for the normal stresses caused by bending at a section is given as

$$\sigma = \frac{My}{I} \tag{5.1}$$

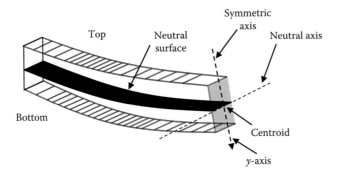

Figure 5.2 Bending of a beam.

where

    σ = normal stress at an arbitrary point on the section

    M = bending moment acting on the section

    y = distance from the neutral axis to the arbitrary point with positive y for points within the area in tension

    I = second moment of area of cross section, representing a geometric quantity of the cross section and having a unit of, for example, m⁴. The mathematical expression of I is

$$I = \int_A y^2 dA(rea) \tag{5.2}$$

The second moment of area of cross section is an equivalent geometric quantity to the polar second moment of area, J, in torsion and the cross-sectional area, A, in axial tension and compression. The respective stresses are inversely proportional to the geometrical quantities of cross sections.

The maximum normal stress occurs at the fibres farthest from the neutral axis, that is, at $y = y_{max}$

$$\sigma_{max} = \frac{My_{max}}{I} = \frac{M}{I/y_{max}} = \frac{M}{W} \tag{5.3}$$

where $W = I/y_{max}$ is called elastic modulus of section.

An efficient beam section has a small ratio of cross-sectional area and elastic modulus of section, A/W. Thus, in terms of the normal stress due to bending, an efficient section concentrates as much material as possible away from the neutral axis. This is why I-shaped sections are widely used in practice.

The elastic modulus of section reflects the geometric properties of a section and is different from the elastic modulus of materials, which depends only on material properties.

5.2 CALCULATION OF SECOND MOMENT OF AREA

The second moment of area is defined by the integral of y^2 over the entire cross-sectional area with respect to an axis that is usually the neutral axis. It is constant for a given section. The following procedure can be followed to compute the integral:

1. Find the centroid of the section.
   a. For a section having two axes of symmetry, the centroid lies at the intersection of the two axes.
   b. For a section having only one axis of symmetry, the centroid lies on the axis. For the section shown in Figure 5.3, the y coordinate of the centroid, $\bar{y}$, is given by

$$\bar{y} = \frac{\int_A y dA}{A} \tag{5.4}$$

where x is an arbitrary axis that is perpendicular to the axis of symmetry, y. $\int_A y dA$ is called *first moment of area* and A is the entire area of the section. From Equation 5.4,

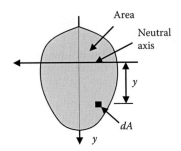

Figure 5.3 Coordinate system for calculating section properties.

it can be seen that when the area and the location of centroid of a section are known, the first moment of area can be easily computed by

$$\int_A y\,dA = \bar{y} \times A \tag{5.5}$$

If A consists of a number of small sub-area(s) A_i and the y coordinates of their centroids are, respectively, $\bar{y}_i$, Equation 5.5 is equivalent to

$$\sum_i A_i \bar{y}_i = \bar{y} \times A \tag{5.6}$$

2. Find the neutral axis.
   a. For a section having two axes of symmetry, the neutral axis is one of them depending on the direction of bending moment applied on the section.

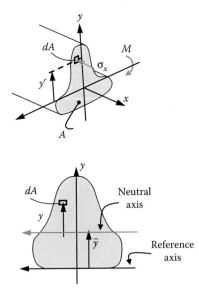

Figure 5.4 Position of neutral axis.

Table 5.1 Second moment of area about neutral axis

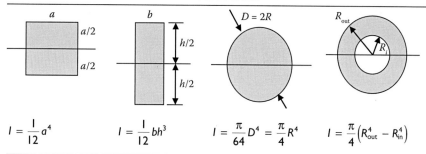

a	b	$D = 2R$	R_{out}
$I = \dfrac{1}{12}a^4$	$I = \dfrac{1}{12}bh^3$	$I = \dfrac{\pi}{64}D^4 = \dfrac{\pi}{4}R^4$	$I = \dfrac{\pi}{4}\left(R_{out}^4 - R_{in}^4\right)$

b. For a section having a single axis of symmetry, the neutral axis is perpendicular to the axis of symmetry and passes though the centroid of the section (Figure 5.4).
3. Calculate second moment of area.
   a. The area integration is usually necessary only for a few regular shapes such as rectangles and circles. Most cross-sectional areas used in practice may be broken into an assembly of these regular shapes. Table 5.1 lists the second moments of area for some of the most commonly used shapes of section.
   b. For a complex cross-sectional area that is an assembly of the above regular shapes, the *parallel axis theorem* is used to compute the second moment of area.

The parallel axis theorem states that the second moment of area of a section, I, about an arbitrary axis equals the second moment of area of the same section about a parallel axis passing through the section's centroid, I_0, plus the product of the area of the section, A, and the square of the distance between the two axes, d, that is,

$$I = I_0 + Ad^2 \tag{5.7}$$

For example, the T section shown in Figure 5.5 is composed of two rectangles. The second moment of area of the entire T section about its neutral axis is obtained by adding the second moments of area of the two rectangles about the same axis, that is

$$I_{NA}(T) = I_{NA}^{(1)}(\bar{\ }) + I_{NA}^{(2)}(|)$$

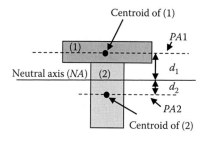

Figure 5.5 Coordinate system for parallel axis theorem.

where

$$I_{NA}^{(1)}(\bar{\ }) = I_{PA1}^{(1)} + A^{(1)}(d_1)^2$$

$$I_{NA}^{(2)}(\mathsf{l}) = I_{PA2}^{(2)} + A^{(2)}(d_2)^2$$

$I_{PA1}^{(1)}$ = Second moment of area (1) about axis $PA1$

$I_{PA2}^{(2)}$ = Second moment of area (2) about axis $PA2$

$A^{(1)}$ = Area of section (1)

$A^{(2)}$ = Area of section (2)

$PA1$ = Parallel axis passing through the centroid of $A^{(1)}$

$PA2$ = Parallel axis passing through the centroid of $A^{(2)}$

5.3 SHEAR STRESSES IN BEAMS

In Figure 5.1d, it can be noted that the shear force on the section is distributed over the same section in the form of shear stresses. It can be conceptually argued that the vertical shear stresses along the top and bottom boundaries of the section must vanish. This is because of the fact that shear stresses always occur in pairs on two perpendicular planes (Figure 5.6). Since there is no shear stress on the top and bottom surfaces of the beam, the vertical shear stresses on the cross section along the intersection of, for example, the top surface and the cross section must also be zero. Thus, the distribution of shear stress across the depth of a beam, τ, can be parabolic (zero along the top and bottom sides and nonzero in between). For prismatic beams subjected to bending, the shear stress is distributed parabolically, depending on the distance to the neutral axis. For symmetric bending, shear stresses are constant along any straight lines that are parallel to the neutral axis

$$\tau = \frac{VS^*}{bI} \tag{5.8}$$

where

V = shear force acting on the section

b = breadth of the beam at the location where the shear stress is computed

I = second moment of area of the entire section about its neutral axis

S^* = first moment of area about the neutral axis for the area enclosed by the boundary and the parallel line passing through the point at which shear stress is computed, that is, for the shaded area of Figure 5.7:

$$S^* = \int_{A^*} y dA^* \tag{5.9}$$

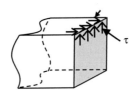

Figure 5.6 Shear stress along the top edge of section.

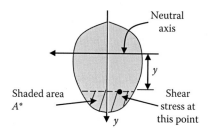

Figure 5.7 Calculation of S*.

5.4 PLASTIC DEFORMATION OF BEAMS

Since Hooke's law was used in deriving Equation 5.1 for computing normal stresses in a beam due to bending, the normal stresses increase linearly as the bending moment increases. However, once a stress exceeds the yield stress of material of the beam (see Figure 1.14), the stress–strain relation becomes nonlinear, in which, in most cases, plastic deformation occurs. In engineering applications, Figure 1.14 can be simplified by the idealised strain–stress diagram shown in Figure 5.8.

It is assumed that the material has the same properties in tension and compression. The strains continue to increase during yielding, without further increase of the stress. Plastic deformation of a section starts from the fibres that have been stressed at or above the yield stress of material, and progressively spreads to other fibres when the bending moment is further increased. Eventually, the entire section will deform plastically. On a cross section of a beam, normally yielding starts from the fibres that are furthest from the neutral axis since the normal stresses of these fibres exceed the yield stress of the material first. At this stage, as shown in Figure 5.9a, the entire section is still elastic and the distribution of bending stresses is linear across the neutral axis of the section. Further increase of the applied bending moment will increase the amount of yielded fibres progressively spreading to the interior zone of the beam section symmetrically from the top

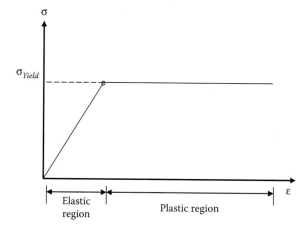

Figure 5.8 Idealised strain–stress curve.

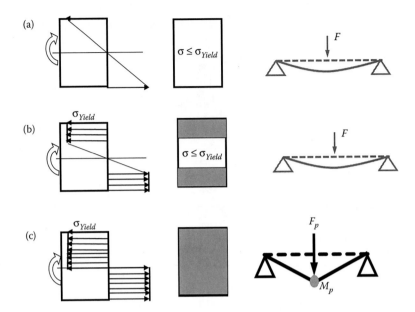

Figure 5.9   (a) Stress caused by F is smaller than σ_{Yield} and the beam deflects elastically σ_{Yield}. (b) Stress caused by F reaches σ_{Yield} on part of the section and a plastic hinge is developing. (c) Stress caused by F reaches σ_{Yield} on the entire section and a plastic hinge has been developed.

and the bottom surfaces. The stress distribution shown in Figure 5.9b applies after a large amount of plastic deformation takes place. The areas subjected to constant yield stress are the plastic zones (shaded area). The ultimate stress distribution is shown in Figure 5.9c, where the whole upper and lower half of the beam are, respectively, subjected to uniform compressive and tensile stress of σ_{Yield}.

There are two important moments developed on a section when the applied loads increase from Figure 5.9a–c, that is,

1. The bending moment of a beam of rectangular section when the furthest fibres just reach σ_{Yield}, as given by Equation 5.3, is called yield moment and can be computed by

$$M_Y = \sigma_{Yield} W = \sigma_{Yield} \frac{bh^2}{6} \tag{5.10}$$

2. The bending moment of a beam of rectangular section when the whole section has reached σ_{Yield}, as shown in Figure 5.9c, is

$$M_p = \sigma_{Yield} \frac{bh^2}{4} \tag{5.11}$$

where M_p is called *plastic moment* of a beam section. It is defined as the maximum moment the section can resist. At this point, a *plastic hinge* (see Figure 5.9c) is formed, which allows large rotations to occur at constant plastic moment M_p. The beam is now a one degree-of-freedom mechanism.

The plastic moment for a given section is always larger than the yield moment. The ratio M_p/M_Y depends only on the cross-sectional properties of a beam and is called the shape factor. The shape factor for the above rectangular section is

$$M_p/M_Y = \frac{6}{4} = 1.5$$

which means that M_Y must be exceeded by 50% before a plastic hinge is formed.

Equation 5.11 can be rewritten in a general form as

$$M_p = \sigma_{Yield}Z \tag{5.12}$$

where Z is called the plastic section modulus. For the rectangular section analysed above $Z = bh^2/4$.

5.5 KEY POINTS REVIEW

- The centroid is the centre of a section.
- The second moment of area represents the geometric contribution of a cross section to the bending resistance.
- The second moment of area of a section about any axis is the second moment of area of the section about a parallel axis passing through the area's centroid, and its area times the square of the distance from the centroid to the axis (the parallel axis theorem).
- A cross-sectional plane remains plane during bending.
- There is no normal (axial) stress or strain at the neutral axis during bending.
- Normal stress is compressive on one side of the neutral axis and tensile on the other side.
- The bending moment on a cross section is the resultant of the normal stresses distributed on the same section.
- Normal strains and stresses are linearly distributed across the depth of a beam, proportional to the distance from the neutral axis.
- If a cross section is made of one material, the normal stress is distributed continuously across the depth.
- Maximum normal stress occurs at a point furthest from the neutral axis.
- The ratio between the second moment of area and the distance from neutral axis to the farthest point is termed *elastic modulus of section* and is a very important design parameter.
- An efficient section in bending concentrates more material away from the neutral axis such that a maximum elastic modulus of section (W) can be achieved with the use of a minimum cross-sectional area (A).
- For materials with different tensile and compressive strength, a shift of the neutral axis from the mid-depth of sections is desirable.
- Shear stress is distributed parabolically across the depth of beam.

- For a section with constant breadth, b, maximum shear stress occurs along the neutral axis.
- For a section with variable breadth, b, maximum shear stress occurs at the location where S^*/b is maximum.
- Shear stress distribution discontinues (abrupt increase or decrease) at the locations where the breadths of section have abrupt changes.
- The shear force on a cross section is equal to the resultant of the shear stresses distributed on the same section.
- On a cross section of a beam, fibres that are furthest from the neutral axis yield first.
- On a cross section, the bending moment associated with the first occurrence of fibre yielding is called yield moment.
- More fibres will yield when the moment on the section is further increased and plastic zone develops progressively, spreading inwards from the outer fibres to the neutral axis.
- Once all the fibres of a section are yielded, the whole section is plastic and a plastic hinge is formed.
- On a cross section, the bending moment associated with a plastic hinge is called the plastic moment.
- Plastic hinge allows excessive rotation, which is equivalent to a pin connection, and may cause total collapse of a structure.
- The ratio of yield moment and plastic moment is called the shape factor.

5.6 RECOMMENDED PROCEDURE OF SOLUTION

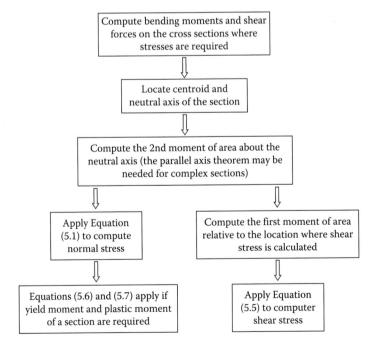

5.7 EXAMPLES

EXAMPLE 5.1

A beam of hollow circular section is loaded with concentrated point loads as shown in figure (a) below. The inside and outside diameters of the hollow circular section are 45 and 60 mm, respectively. Calculate the maximum normal stress of the beam.

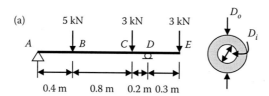

Solution

For a beam with constant cross section, the maximum normal stress and the maximum bending moment always occur on the same cross section. A bending moment diagram is essential to determine the maximum bending moment and its location. This can be done by following the procedure described in Chapter 4.

Step 1: Compute support reactions (figure [b] below)
     Taking anticlockwise moment about A

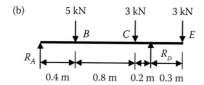

$$R_D \times 1.4 \text{ m} - 5 \text{ kN} \times 0.4 \text{ m} - 3 \text{ kN} \times 1.2 \text{ m} - 3 \text{ kN} \times 1.7 \text{ m} = 0$$
$$R_D = 7.64 \text{ kN } (\uparrow)$$

Resolving vertically

$$R_A + R_D - 5 \text{ kN} - 3 \text{ kN} - 3 \text{ kN} = 0$$
$$R_A = 11 \text{ kN} - V_D = 3.36 \text{ kN } (\uparrow)$$

Step 2: Compute the bending moments on the critical sections at A, B, C, D and F (figure [c] below)

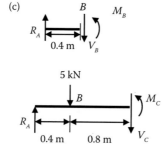

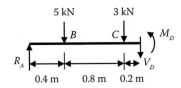

1. Section at A
   Due to the pin support

   $$M_A = 0$$

2. Section at B
   Taking anticlockwise moment about B

   $$M_B - R_A \times 0.4 \text{ m} = 0$$

   $$M_B = 3.36 \text{ kN} \times 0.4 \text{ m} = 1.34 \text{ kN m } (\circlearrowleft)$$

3. Section at C
   Taking anticlockwise moment about C

   $$M_C - R_A \times 1.2 \text{ m} + 5 \text{ kN} \times 0.8 \text{ m} = 0$$
   $$M_C = 3.36 \text{ kN} \times 1.2 \text{ m} - 5 \text{ kN} \times 0.8 \text{ m} = 0.032 \text{ kN m } (\circlearrowleft)$$

4. Section at D
   Taking anticlockwise moment about D

   $$M_D - R_A \times 1.2 \text{ m} + 5 \text{ kN} \times 1.0 \text{ m} + 3 \text{ kN} \times 0.2 \text{ m} = 0$$
   $$M_D = 3.36 \text{ kN} \times 1.4 \text{ m} - 5 \text{ kN} \times 1.0 \text{ m} - 3 \text{ kN} \times 0.2 \text{ m}$$
   $$= -0.896 \text{ kN m } (\circlearrowright)$$

5. Section at E
   Due to the free end

   $$M_E = 0$$

Step 3: Draw the bending moment diagram (figure [d] below)
Since there is no distributed load applied between the critical sections, the bending moment distributions between these sections are slopping lines.

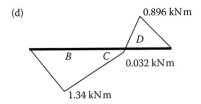

(d)

It is clear from the bending moment diagram that the maximum magnitude of bending moment occurs on the section at B, that is,

$$M_{max} = 1.34 \text{ kN m}$$

Thus, the maximum normal stress occurs on the same section.

Step 4: Compute second moment of area

From Table 5.1, the second moment of area of the hollow circular section

$$l = \frac{\pi}{64}(R_{out}^4 - R_{in}^4) = \frac{\pi}{64}(D_{out}^4 - D_{in}^4) = \frac{\pi}{64}(60^4 - 45^4) \text{ mm}^4$$

$$= 4.35 \times 10^{-7} \text{ m}^4$$

Step 5: Compute normal stress

From Equations 5.1 and 5.2, at *B*

$$\sigma_{max} = \frac{M_{max}y_{max}}{I} = \frac{M_{max}D_{out}/2}{I} = \frac{1.34 \text{ kN m} \times 30 \times 10^{-3} \text{ m}}{4.35 \times 10^{-7} \text{ m}^4}$$

$$= 92.41 \text{ MN/m}^2$$

EXAMPLE 5.2

Simply supported steel beam is subjected to a uniformly distributed load as shown in figure below. The maximum allowable normal stress of the material is 160 MPa. Design the beam using a circular section, a rectangular section of $h/b = 2$ and an I-shaped section, respectively.

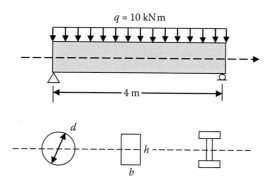

Solution

The maximum bending moment occurs at the mid-span. The maximum normal stress on the cross section at the mid-span must be calculated first. The stress is then compared with the allowable stress of the material to determine the size of the cross section.

Step 1: Compute the reaction forces at the supports

Since the beam is symmetrically loaded, the two support reactions are equal to half of the total applied load and act vertically upwards.

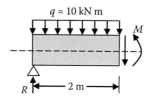

$$R = \frac{1}{2}(10 \times 4) = 20 \text{ kN}$$

Step 2: Compute the mid-span (maximum) bending moment

$$M = 2R - 10 \times 2 \times 1 = 40 - 20 = 20 \text{ kN m}$$

Step 3: Design selections

The maximum normal stress on the section at the mid-span is

$$\sigma = \frac{M y_{max}}{I} = \frac{M}{I/y_{max}} = \frac{M}{W}$$

1. For a circular section

$$I = \frac{\pi d^4}{64}$$

$$y_{max} = \frac{d}{2}$$

$$W = \frac{I}{y_{max}} = \frac{\pi d^3}{32}$$

Thus,

$$\sigma = \frac{M}{I/y_{max}} = \frac{20}{\pi d^3/32} \leq 160 \times 10^6$$

$$d \geq \sqrt[3]{\frac{20 \times 10^3 \times 32}{\pi \times 160 \times 10^6}} = 10.838 \times 10^{-2} \text{ m} = 10.84 \text{ cm}$$

The minimum area of the circular section

$$A_C = \frac{\pi d^2}{4} = \frac{\pi \times 10.84^2}{4} = 92.29 \text{ cm}^2$$

2. For a rectangular section ($h = 2b$)

$$I = \frac{1}{12} bh^3 = \frac{1}{12} b(2b)^3 = \frac{2}{3} b^4$$

$$y_{max} = \frac{h}{2} = b$$

Thus,

$$\sigma = \frac{M}{I/y_{max}} = \frac{20}{2b^3/3} \leq 160 \times 10^6$$

$$b \geq \sqrt[3]{\frac{3 \times 20 \times 10^3}{2 \times 160 \times 10^6}} = 5.724 \times 10^{-2} \text{ m} = 5.72 \text{ cm}$$

The minimum area of the rectangular section

$$A_R = b \times h = 2b^2 = 2 \times 5.72^2 = 65.44 \text{ cm}^2$$

3. For an I-shaped section (UB section)
From the British Standard (BS) sections, I/y_{max} is named as *elastic modulus of a section*, which is used as a design parameter:

$$\sigma = \frac{M}{I/y_{max}} = \frac{20}{I/y_{max}} \leq 160 \times 10^6$$

$$I/y_{max} \geq \frac{20 \times 10^3}{160 \times 10^6} = 0.125 \times 10^{-3}\,\text{m}^3 = 125\,\text{cm}^3$$

The BS 5950 provides a table of elastic moduli sections for many UB (I-shaped) sections. The modulus that is immediately larger than 125 cm³ from the list is 153 cm³. The section designation of this modulus is UB $178 \times 102 \times 19$ that has a cross-sectional area of 24.3 cm². Thus, the minimum area of the I-shaped section is

$$A_I = 24.3\,\text{cm}^2$$

The comparison for the weights of the beam using the three different sections can be made by comparing the cross-sectional areas of the sections. Obviously, among the three sections, the I-shaped section is the most economic one, which is more than three times lighter than the circular section.

EXAMPLE 5.3

A beam having the cross section as shown in figure (a) below is subjected to a shear force of 12 kN and a bending moment of 12 kN m on a section. Compute (a) the normal stresses along the top and bottom surfaces of the hollow area; (b) the maximum magnitudes of the normal and shear stresses and (c) the normal and shear stress distributions.

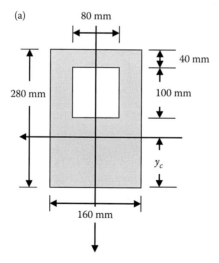

Solution

The cross section is not symmetric about any horizontal axis. Thus, the vertical location of the centroid must be found first. The horizontal axis passing through the centroid is then the neutral axis. This can be done by treating the area enclosed by the outside boundary and deducting from it the hollow area enclosed by the inside boundary.

Step 1: Compute the vertical distance of the centroid from the bottom side
This can be done in the following tabular form.

Area	A(mm^2)	$\bar{y}$ (mm) from the bottom	$\int y\,dA = A\bar{y}$ (mm^3)
Gross section	$160 \times 280 = 44{,}800$	140	6,272,000
Hollow area	$100 \times 80 = 8000$	$140 + 50 = 190$	1,520,000

Thus, from Equations 5.4 to 5.6 for the actual section

$$\sum A = 44{,}800 - 8000 = 36{,}800 \text{ mm}^2$$

$$\sum A\bar{y} = 6{,}272{,}000 - 1{,}520{,}000 = 4{,}752{,}000 \text{ mm}^3$$

$$y_C = \frac{\sum A\bar{y}}{\sum A} = \frac{4{,}752{,}000}{36{,}800} = 129.1 \text{ mm}$$

The neutral axis is 129.1 mm above and parallel to the bottom side of the section.

Step 2: Compute the second moment of area from the parallel axis theorem
For the gross area enclosed by the outside boundary (figure [b] below), the second moment of area about the neutral axis is computed from Equation 5.7 as

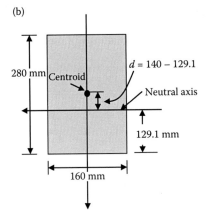

(b)

$$I_{\text{gross}} = I_0 + Ad^2 = \frac{160 \times 280^3}{12}$$
$$+ 280 \times 160 \times (140 - 129.1)^2$$
$$= 2.98 \times 10^8 \text{ mm}^4$$

For the hollow area (figure [c] below), the second moment of area about the neutral axis is also computed from Equation 5.7 as

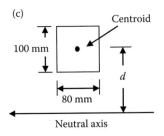

(c)

$$I_{\text{hollow}} = I_o + Ad^2 = \frac{90 \times 100^3}{12}$$

$$+ 100 \times 80 \times (190 - 129.1)^2$$

$$= 3.72 \times 10^7 \text{ mm}^4$$

The second moment of area of the cross-sectional area

$$I = I_{\text{gross}} - I_{\text{hollow}}$$

$$= (29.8 - 3.72) \times 10^7 = 26.1 \times 10^7 \text{ mm}^4$$

Step 3: Compute normal stresses
  Along the top surface of the hollow area
    The distance from the neutral axis to the surface, that is, the y coordinate of the surface

$$y = -(140 - 129.1 + 100) \text{ mm} = -110.9 \text{ mm}$$

$$\sigma = \frac{My}{I} = \frac{12 \text{ kN m} \times (-110.9) \times 10^{-3} \text{ m}}{26.1 \times 10^{-5} \text{ m}^4} = -5.1 \text{ MN/m}^2 \text{ (compression)}$$

  Along the bottom surface of the hollow area
    The distance from the neutral axis to the surface, that is, the y coordinate of the surface

$$y = -(140 - 129.1) \text{ mm} = -10.9 \text{ mm}$$

$$\sigma = \frac{My}{I} = \frac{12 \text{ kN m} \times (-10.9) \times 10^{-3} \text{ m}}{26.1 \times 10^{-5} \text{ m}^4} = -0.5 \text{ MN/m}^2 \text{ (compression)}$$

Step 4: Compute the maximum magnitude of normal stress
  It is obvious from Equations 5.1 and 5.2 that the maximum magnitude of normal stress occurs along the top side of the section since its distance to the neutral axis is the maximum. This distance is

$$y = -(280 - 129.1) \text{ mm}$$

$$\sigma = \frac{My}{I} = \frac{12 \text{ kN m} \times (-150.9) \times 10^{-3} \text{ m}}{26.1 \times 10^{-5} \text{ m}^4} = -6.94 \text{ MN/m}^2 \text{ (compression)}$$

Step 5: Compute the maximum magnitude of shear stress
  From Equations 5.8 and 5.9, the maximum magnitude of shear stress may occur at the location where (1) the S^* defined in Equations 5.8 and 5.9 is maximum or (2) the breadth of the cross section, b, is minimum.

  1. S^* is always maximum along the neutral axis. Consider the area below the neutral axis:

$$S^* = \sum A\bar{y} = 160 \text{ mm} \times 129.1 \text{ mm} \times \frac{129.1 \text{ mm}}{2} = 1.33 \times 10^{-3} \text{ m}^3$$

$$\tau = \frac{VS^*}{bI} = \frac{12 \text{ kN} \times 1.33 \times 10^{-3} \text{ m}^3}{160 \times 10^{-3} \text{ m} \times 26.1 \times 10^{-5} \text{ m}^4} = 0.382 \text{ MN/m}^2$$

2. b is minimum between the top and the bottom surfaces of the hollow area, while with this range the S^* taken for the area immediately above the bottom side of the hollow part is the largest. Thus, at this location (140 mm from the top)

$$S^* = \sum A\bar{y} = 140 \times 160 \times (140 - 129.1 + 70) - 100 \times 80 \times (140 - 129.1 + 50)$$

$$= 1.32 \times 10^{-3} \text{ m}^3$$

$$\tau = \frac{VS^*}{bI} = \frac{12 \text{ kN} \times 1.32 \times 10^{-3} \text{ m}^3}{2 \times 40 \times 10^{-3} \text{ m} \times 26.1 \times 10^{-5} \text{ m}^4} = 0.76 \text{ MN/m}^2$$

Therefore, the maximum magnitude of shear stress occurs at the location 140 mm below the top surface of the cross section. At the same location, if we take the breadth of the solid part as b in Equations 5.8 and 5.9, that is,

$$\tau = \frac{VS^*}{bI} = \frac{12 \text{ kN} \times 1.32 \times 10^{-3} \text{ m}^3}{160 \times 10^{-3} \text{ m} \times 26.1 \times 10^{-5} \text{ m}^4} = 0.379 \text{ MN/m}^2$$

Observation: Due to the abrupt change of the breadth of section when across the bottom side of the hollow area (from 160 to 80 mm), the shear stress jumps from 0.379 to 0.76 MN/m². It can be concluded that on a cross section if the breadth of section changes suddenly across a line, the shear stress also has a sudden change (increased or decreased) by the ratio of change in breadth, that is, the ratio of the breadth immediately below and above the line. In this example, the ratio is 160/80 = 2. The shear stress above the line (140 mm from the top) is therefore twice as big as it is below the line.

By calculating the normal stress at the bottom surface and the shear stresses above and below the top side of the hollow area, the normal and shear stress distribution on the cross section can be sketched as below.

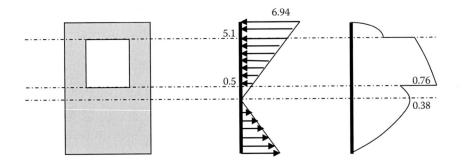

Obviously, a sudden change in geometry of the cross section does not introduce sudden change in normal stress, but in shear stress distribution.

EXAMPLE 5.4

Determine the maximum normal stresses in the concrete and the steel in a simply supported reinforced concrete beam subjected to uniformly distributed load of 1250 lb/ft over a span of 25 ft. The cross section of the beam, as shown in figure below, is reinforced with three steel bars having a total cross-sectional area of π in.². Assume that the ratio of E for steel to that for concrete is 15 and ignore all concrete in tension.

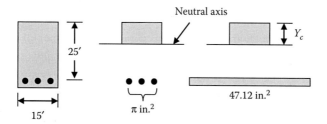

Solution

Since the stress-carrying capacity of concrete in tension zone is significantly smaller compared with steel, all concrete below the neutral axis is neglected. The bending moment on a cross section of the beam is balanced by the compression in the concrete and the tension in the steel reinforcement. To compute the location of the neutral axis and stress distribution on the cross section, the area of the steel needs to be transferred to its equivalent area of concrete on the basis of the ratio of E, so that all the formulas developed for beam-bending problems are valid. The transformation is based on the assumption that the strains and the resultant force in the steel reinforcement and the equivalent concrete are equal.

Step 1: Compute support reactions
 Since the beam is symmetrically loaded, the two support reactions are both equal to half of the total applied load and act vertically upwards

$$R = \frac{1250 \times 25}{2}\,\text{lb}$$

Step 2: Compute bending moment
 The maximum normal stress occurs on the cross section subjected to the maximum bending. It is obvious that the maximum bending moment occurs at the mid-span. The bending moment at the mid-span is calculated as follows:
 Taking moment about the mid-span

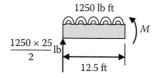

$$M = \frac{1250 \times 25}{2} \times 12.5 - 1250 \times 12.5 \times \frac{12.5}{2}$$
$$= 97,656.25 \,\text{lb ft}$$
$$= 1,171,875 \,\text{lb ft}$$

Step 3: Compute the equivalent concrete area A_{eq} of the steel
 Assume that on the equivalent concrete A_{eq} and the steel bars A_{steel}, both the strains and the resistance forces are the same. Thus,

$$2\sigma_{steel}A_{steel} = \sigma_{eq}A_{eq} \tag{5.13}$$

or

$$E_{steel}\varepsilon_{steel}A_{steel} = E_C\varepsilon_{eq}A_{eq}$$

Since

$$\varepsilon_{steel} = \varepsilon_{eq}$$

$$A_{eq} = \frac{E_{steel}}{E_c}A_{steel} = 15A_{steel} = 15\pi = 47.12 \text{ in.}^2$$

(5.14)

Thus, the steel reinforcement can be replaced by an equivalent concrete area of 47.12 in.² (see Example 5.4 figure).

Step 4: Compute vertical distance of the centroid from the top

Area	$A(\text{in.}^2)$	$\bar{y}$ (in.) from the top	$\int y dA = A\bar{y}$ (in.³)
Concrete in compression	$15 \times y_c$	$y_c/2$	$7.5y_c^2$
Equivalent concrete	47.12	25	1178

From Equations 5.4 to 5.6

$$\sum A = (15y_c + 47.12)\text{ in.}^2$$
$$\sum A\bar{y} = (7.5y_c^2 + 1178)\text{ in.}^3$$
$$y_c = \frac{\sum A\bar{y}}{\sum A} = \frac{7.5y_c^2 + 1178}{15y_c + 47.12}$$

Hence,

$$7.5y_c^2 + 47.12y_c - 1178 = 0$$
$$y_c = 9.4 \text{ in.}$$

The neutral axis is 9.4 in. below the top surface.

Step 5: Compute second moment of area of the transferred cross section. For the concrete (Example 5.4 figure)

$$I_c = \frac{1}{12} \times 15 \times y_c^3 + 15 \times y_c \times \left(\frac{y_c}{2}\right)^2$$

$$= \frac{1}{12} \times 15 \times 9.4^3 + 15 \times 9.4 \times \left(\frac{9.4}{2}\right)^2$$

$$= 4152.92 \text{ in.}^4$$

For the equivalent concrete area (Example 5.4 figure)

$$I_{eq} = 0 + A_{eq} \times (25 - y_c)^2 = 11,467.12 \text{ in.}^4$$

For the entire section

$$I = I_c + I_{eq} = 15,620.04 \text{ in.}^4$$

Note that zero is given to the second moment of area of the equivalent area of reinforcement about the parallel axis passing through its own centroid. This is because the thickness of the equivalent concrete area is very small and the resulting second moment of area is far smaller than the second term in I_{eq}.

Step 6: Compute normal stresses
Along the top, the maximum compressive stress in concrete occurs

$$\sigma = \frac{My}{I} = \frac{1,171,875 \times 9.4}{15,620.04} = 705.2 \text{ lb/in.}^2 \text{ (compression)}$$

On the equivalent area of reinforcement, the maximum tensile stress occurs

$$\sigma_{eq} = \frac{My}{I} = \frac{1,171,875 \times (25 - 9.4)}{15,620.04} = 1170.4 \text{ lb/in.}^2 \text{ (tension)}$$

The actual tensile stress in the steel can be computed from Equations 5.13 and 5.14, that is,

$$\sigma_{steel} = \frac{A_{eq}}{A_{steel}} \sigma_{eq} = \frac{E_{steel}}{E_c} \sigma_{eq} = 15 \times 1170.4 = 17,555.6 \text{ lb/in.}^2$$

EXAMPLE 5.5

Derive an expression for the shear stress distribution on a rectangular section subjected to a shear force V (figure [a] below).

(a)

Solution

This question asks for a straightforward application of Equations 5.8 and 5.9. In order to find the expression of shear stress distribution, shear stress at an arbitrary location must be calculated. The cross section is symmetric and the horizontal axis of symmetry is the neutral axis of the section.

Step 1: Select an arbitrary point on the cross section. The vertical distance from the point to the neutral axis is y ($0 \le y \le h/2$) (figure [b] below).

(b)

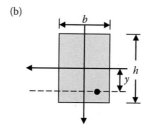

Step 2: Draw a line passing through the selected point and parallel to the neutral axis (figure below).

Step 3: Calculate the first moment of area S* (Equations 5.8 and 5.9). The parallel line separates the cross section into two parts, one of which *does not* include the neutral axis (figure [c] below). Usually the first moment of area of this part about the neutral axis is calculated as S*.

(c)

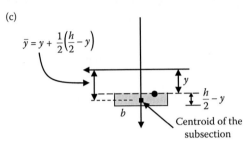

$$\bar{y} = y + \frac{1}{2}\left(\frac{h}{2} - y\right)$$

Centroid of the subsection

The area of the part

$$A^* = b \times \left(\frac{h}{2} - y\right)$$

The distance from the neutral axis to the centroid of the part

$$\bar{y} = y + \frac{1}{2}\left(\frac{h}{2} - y\right) = \frac{h}{4} + \frac{y}{2}$$

The first moment of area of the part about the neutral axis

$$S^* = A^*\bar{y} = b\left(\frac{h}{2} - y\right) \times \left(\frac{h}{4} + \frac{y}{2}\right)$$

$$= \frac{b}{2}\left[\left(\frac{h^2}{4} - y^2\right)\right]$$

Step 4: Compute shear stress distribution

The second moment of area of the section about its neutral axis is

$$I = \frac{1}{12}bh^3$$

and the breadth of the section at the arbitrary location is b. Thus,

$$\tau = \frac{VS^*}{bI} = \frac{12V}{b(bh^3)} \times \frac{b}{2}\left[\left(\frac{h^2}{4} - y^2\right)\right]$$

$$= \frac{6V}{bh^3}\left[\left(\frac{h^4}{4} - y^2\right)\right]$$

The distribution of the shear stress is a parabolic function of the distance, y, from the neutral axis as shown in figure (d) below. It is obvious that the maximum shear stress occurs along the neutral axis, that is, at y = 0.

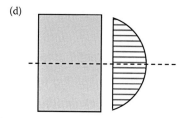

(d)

From the expression of distribution

$$\tau_{max} = \frac{6V}{bh^3}\left[\left(\frac{h^2}{4} - y^2\right)\right]_{y=0} = \frac{3V}{2bh} = \frac{3V}{2A}$$

where $A(= bh)$ is the cross-sectional area of the beam. V/A represents the average shear stress on the section.

• The maximum shear stress due to bending of a rectangular section is 1.5 times larger than the average shear stress on the section and occurs along the neutral axis.

EXAMPLE 5.6

The compound beam shown in figure (a) below is composed of two identical beams of rectangular section: (a) if the two beams are simply placed together with frictionless contact and the maximum allowable normal stress of the material is $[\sigma]$, calculate the maximum value of the force, P, that can be applied on the beam; (b) if the two beams are fastened together by a bolt as shown in figure (b) below, what is the maximum value of P; and (c) if the maximum allowable shear stress in the bolt is $[\tau]$, calculate the minimum diameter of the bolt when the compound beam is loaded with the P calculated from (b).

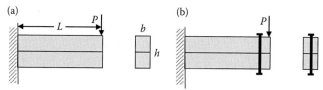

Solution

This example tests your understanding of bending deformation in relation to neutral axis. In case (a), due to the frictionless contact, the two beams slide on each other and deform independently, each of which carries a half of the total bending moment induced by P. Therefore, the two beams have their own neutral axes. In case (b), the two beams are fastened together so that the compound beam acts as a unit and has only one neutral axis.

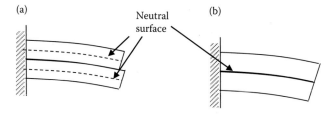

1. The maximum normal stress occurs at the fixed end where the bending moment is the maximum:

$$M_{max} = PL$$

The maximum bending moment carried by each of the beams is therefore $PL/2$. Thus, by Equations 5.1 and 5.2, the maximum normal stress within each of the beams is

$$\sigma = \frac{My}{I} = \frac{(PL/2) \times (h/4)}{(1/12) \times b \times (h/2)^3} = \frac{12\,PL}{bh^2} \leq [\sigma]$$

So, the maximum force that can be applied on the beam is

$$P_{max} \leq \frac{[\sigma]bh^2}{12L}$$

2. When the two beams act as a unit, the contact surface is now the neutral surface. The maximum normal stress is then

$$\sigma = \frac{My}{I} = \frac{PL \times (h/2)}{(1/12) \times b \times h^3} = \frac{6PL}{bh^2} \leq [\sigma]$$

So the maximum force that can be applied on the beam for this case is

$$P_{max} \leq \frac{[\sigma]bh^2}{6L}$$

It can be seen that the bonded beam has higher load-carrying capacity than that of the two separated beams.

3. Assume that the two beams are bonded perfectly together, and the maximum shear stress occurs along the neutral axis. Thus, by Equation 5.7

$$\tau_{max} = \frac{VS^*}{bI} = \frac{P_{max}}{b \times (1/12)bh^3} \times \left(\frac{h}{2}\right) \times \left(\frac{h}{4}\right)$$

$$= \frac{[\sigma]bh^2}{6L} = \frac{3}{2bh} = \frac{[\sigma]h}{4L}$$

Since the beam has a constant shear force distribution along the axis, this maximum shear stress applies to any cross section of the beam. Thus, the resultant of the shear stresses acting on the neutral surface is

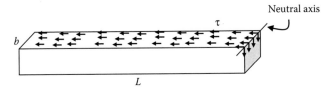

$$Q = \tau_{max} \times b \times L$$

In case (c), the shear force is entirely carried by the bolt that has a cross-sectional area of $\pi d^2/4$, where d is the diameter of the bolt. The shear stress in the bolt is

$$\tau = \frac{Q}{A} = \frac{\tau_{max}bL}{\pi d^2/4} = \frac{[\sigma]h}{4L} \times \frac{bL}{\pi d^2/4} = \frac{[\sigma]hb}{\pi d^2} \leq [\tau]$$

So, the minimum diameter of the bolt is

$$d \geq \sqrt{\frac{[\sigma]bh}{\pi[\tau]}}$$

EXAMPLE 5.7

Determine the elastic–plastic moment of the cantilever shown in figure below. Assume that the beams are made of an ideal elastic–plastic material.

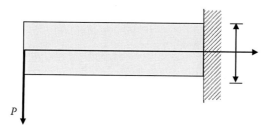

Solution

We shall calculate the moment by deriving the stress distribution at an arbitrary loading stage at which part of the cross section has yielded and is plastic. We shall study the moment at the fixed end where yielding starts first.

Step 1: Assume that P has caused yielding on the cross section at the fixed end and the section shows distinctive zones of elastic and plastic deformation (shaded areas), as shown below.

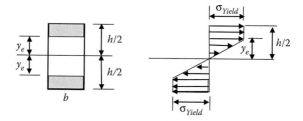

Step 2: The resultant bending moment of the elastic and plastic stresses shown above can be calculated as follows:

- The resultant forces acting on the plastic zones are, respectively,

$$F_p = \pm \sigma_{Yield} b \left(\frac{h}{2} - y_e \right)$$

- The distance from the centroid of the plastic zone to the neutral axis is

$$d_p = y_e + \frac{1}{2}\left(\frac{h}{2} - y_e \right) = \frac{1}{2}\left(\frac{h}{2} + y_e \right)$$

- The resultant forces acting on the elastic zones are, respectively,

$$F_e = \pm \frac{1}{2}\sigma_{Yield} b y_e$$

- The distance from the centroid of the elastic zone to the neutral axis is

$$d_e = \frac{2}{3} y_e$$

Thus, the resultant bending moment of all the stresses on the section is

$$M = 2(F_p d_p + F_e d_e) = \sigma_{Yield} \frac{bh^2}{4} - \frac{1}{3} \sigma_{Yield} b y_e^3$$

From Equation 5.11, the moment can be written as

$$M = M_p - \frac{1}{3} \sigma_{Yield} b y_e^2$$

One the one hand, it is clear from the above solution that when the whole section is plastic, which means when $y_e = 0$,

$$M = M_p$$

On the other hand, if $y_e = h/2$,

$$M = \sigma_{Yield} \frac{bh^2}{4} - \frac{1}{3} \sigma_{Yield} b \left(\frac{h}{2}\right)^2 = \sigma_{Yield} \frac{bh^2}{6}$$

which is the yield moment shown in Equation 5.10.

EXAMPLE 5.8

If the cross section of the cantilever shown in figure above is a solid circle of diameter D, calculate the plastic moment of the section.

Solution

To find the plastic moment of the section, we assume that the whole section is plastic and all the fibres has reached the yield stress, as shown below.

Step 1: Stress distribution on the yield section

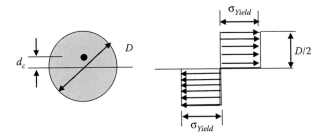

Step 2:

- The resultant forces of the tensile and compressive stress are, respectively,

$$F_p = \pm \sigma_{Yield} \frac{\pi D^2}{8}$$

- The distance from the centroid of the tension or compression zone is

$$d_c = \frac{4D}{6\pi}$$

Thus, the resultant bending moment of all the stresses on the section is

$$M_p = 2(F_p d_c) = 2\sigma_{Yield}\frac{\pi D^2}{8}\frac{4D}{6\pi} = \sigma_{Yield}\frac{D^3}{6}$$

From Equation 5.12, the plastic section module of a solid circular section is $Z = D^3/6$.

5.8 CONCEPTUAL QUESTIONS

1. What is meant by 'neutral axis'?
2. What is meant by 'the second moment of area of a cross section'? If this quantity is increased, what is the consequence?
3. What is the parallel axis theorem, and when can it be used?
4. What is meant by 'elastic modulus of section'? Explain how it is used in steel design.
5. How is the normal stress distributed on a beam section during bending?
6. Why do normal stresses vary across a beam's cross section during bending?
7. When a beam is under bending, the magnitudes of the maximum compressive and the maximum tensile stresses are always the same? (Y/N)
8. The cantilever shown in Figure 5.8 has an I-shaped section and is loaded with a uniformly distributed pressure. The dashed line passes through the centroid of the cross section.

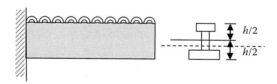

Which of the following statements is correct?
   a. The dashed line is the neutral axis and the maximum tensile stress occurs in the fibre along the bottom surface of the cross section at the fixed end.
   b. The dashed line is the neutral axis and the maximum tensile stress occurs in the fibre along the top surface of the cross section at the fixed end.
   c. The solid line is the neutral axis and the maximum tensile stress occurs in the fibre along the bottom surface of the cross section at the fixed end.
   d. The solid line is the neutral axis and the maximum tensile stress occurs in the fibre along the bottom surface of the cross section at the fixed end.
9. Explain why symmetric sections are preferable for beams made of materials with equal tensile and compressive strengths, while unsymmetrical sections are preferable for beams made of materials with different tensile and compressive strengths.
10. When a beam is under bending, the maximum shear stress always occurs along the neutral axis? (Y/N)

11. The cross section shown in figure below is loaded with a shear force V. Which one of the following statements is correct when Equations 5.8 and 5.9 are used to calculate the shear stress along $m-m$?

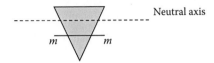

a. In Equations 5.8 and 5.9, S^* denotes the first moment of area of the entire cross-sectional area about the neutral axis and b is the width of the section along $m-m$.

b. In Equations 5.8 and 5.9, S^* denotes the first moment of area of the entire cross-sectional area about the neutral axis and b is the width of the section along the neutral axis.

c. In Equations 5.8 and 5.9, S^* denotes the first moment of area of the section below $m-m$ about the neutral axis and b is the width of the section along $m-m$.

d. In Equations 5.8 and 5.9, S^* denotes the first moment of area of the section below $m-m$ about the neutral axis and b is the width of the section along the neutral axis.

12. Two beams of the same material and cross section are glued together to form a combined section as shown in figure below.

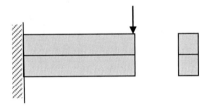

If the beam is subjected to bending, which form of the following normal stress distributions is correct?

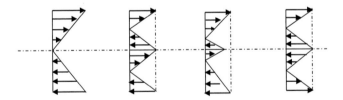

If the two beams are simply placed together with frictionless contact, which form of the above normal stress distributions is correct?

13. In Question 12, if the cross sections of the two beams are channel shaped, as shown in figure below, and the contact between the two beams are frictionless, which form of the normal stress distributions shown in Question 12 is correct?

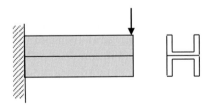

14. The beam shown in figure below is composed of two identical planks that are placed either horizontally or vertically and not fastened together. If the beam is under pure bending, which one of the following statements is correct?

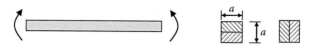

 a. The maximum normal stresses in sections (a) and (b) are the same.
 b. The maximum normal stress in section (a) is greater.
 c. The maximum normal stress in section (b) is greater.
 d. Any of the above statements can be correct, depending on the magnitude of the bending moment applied.

15. On the cross section of a steel beam under bending, which one of the following statements is correct?
 a. At the point where the maximum normal stress occurs, shear stress is always zero, while at the point where the maximum shear stress occurs, normal stress is not necessarily zero.
 b. At the point where the maximum normal stress occurs, shear stress is not zero, while at the point where the maximum shear stress occurs, normal stress is always zero.
 c. At the point where the maximum normal stress occurs, shear stress is always zero, and at the point where the maximum shear stress occurs, normal stress is also zero.
 d. At the point where the maximum normal stress occurs, shear stress is not zero, and at the point where the maximum shear stress occurs, normal stress is also not zero.

16. What is meant by 'yield moment' of a beam section?
17. What is meant by 'plastic moment' of a section?
18. At a plastic hinge of a beam subjected to bending, which one of the following statement is not correct?
 a. The whole section of the beam at the plastic hinge has yielded.
 b. The tensile stresses on the section are equal to or exceed the yield stress.
 c. The compressive stresses on the section are equal to or exceed the yield stress.
 d. The bending moment on this section is equal to zero.

19. Can a beam section carry a bending moment that is greater than the yield moment?
20. Can a beam section carry a bending moment that is greater than the plastic moment?

5.9 MINI TEST

PROBLEM 5.1

The four cross sections shown in figure below have the same cross-sectional area and are made of the same material. Which one of them is the most efficient section and which of them is the least efficient? Explain why.

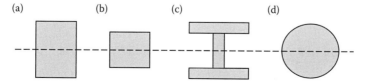

PROBLEM 5.2

Sketch shear stress distributions on the cross sections shown in figure below when they are subjected to a vertical shear force.

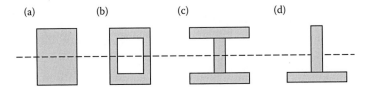

PROBLEM 5.3

A T beam shown in figure below is subjected to a uniformly distributed load. Determine the location of the cross section on which maximum tensile stress and maximum compressive stress occur and calculate the magnitudes of these stresses. Plot the distribution of the normal stress on the cross section.

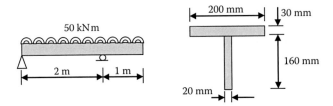

PROBLEM 5.4

For the beam shown in figure above, determine also the location of the cross section on which maximum shear stress occurs. Find an expression for the shear stress distribution on the section and calculate the magnitude of the maximum shear stress.

PROBLEM 5.5

Determine the maximum allowable bending moment that can be applied to the composite beam section shown in figure below. The wood beam is longitudinally reinforced with steel strips on both the top and the bottom sides. The two materials are fastened together so that they act as a unit. The maximum allowable normal stresses of wood and steel are, respectively, 8.3 and 140 MN/m^2 (E_{steel} = 200 GPa, E_{wood} = 8.3 GPa).

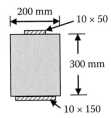

Chapter 6

Deflection of beams under bending

A beam is any long structural member on which loads act perpendicular to the longitudinal axis. If the cross section of the beam has a plane of symmetry and the transversely applied loads are applied within the plane, the axis of the beam will deflect from its original position within the plane of symmetry, as shown in Figure 6.1.

The shape of the deflection curve will depend on several factors, including

- The material properties of the beam as measured by the elastic modulus of material
- The beam's cross section as measured by its second moment of area
- The load on the beam, described as a function of the position along the beam
- The supports of the beam

If the beam is placed in the x–y coordinate system as shown in Figure 6.1, the vertical deflection of the axis, y, is a function of the x coordinate and $\theta(x)$ denotes the slope of the deflection curve at an arbitrary x coordinate, that is, the angle of rotation of the cross section at x. Bending of a beam is measured by curvature of the deflected axis that can be approximately calculated by d^2y/dx^2. The deflection of a beam is characterised by the following:

- Deflection (positive downwards) = $y(x)$.
- Slope of deflection curve = $dy(x)/dx$.
- Curvature of deflection curve = $d^2y(x)/dx^2$.
- Flexural rigidity = EI, representing the stiffness of a beam against deflection. For a given bending moment and at a given section, a stiffer (greater) EI results in a smaller curvature.

6.1 SIGN CONVENTION

A positive bending moment is defined as a moment that induces *sagging* deflection. As shown in Figure 6.2, a positive moment always produces a negative curvature in the adopted coordinate system, where sagging deflection is defined as positive.

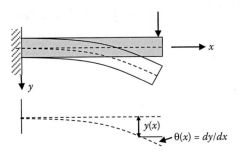

Figure 6.1 Deflection of cantilever.

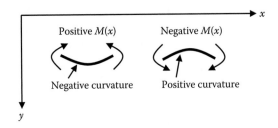

Figure 6.2 Sign convention of bending.

6.2 EQUATION OF BEAM DEFLECTION

The extent that a beam is bent, that is, the curvature of deflection curve, is directly proportional to the applied bending moment $M(x)$, and is inversely proportional to the flexural rigidity EI:

$$\frac{d^2y(x)}{dx^2} = -\frac{M(x)}{EI} \tag{6.1}$$

For a beam with a uniform cross section, from Equations 4.1 and 4.2

$$\frac{d^3y(x)}{dx^3} = -\frac{1}{EI}\frac{dM(x)}{dx} = -\frac{V(x)}{EI} \tag{6.2}$$

$$\frac{d^4y(x)}{dx^4} = -\frac{1}{EI}\frac{dV(x)}{dx} = \frac{q(x)}{EI} \tag{6.3}$$

The negative sign on the right-hand side of Equation 6.1 reflects the sign difference between bending moment and the resulting curvature, following the sign convention described in Section 6.1. The deflection, $y(x)$, can be solved from the above differential equations by various methods of solution.

6.2.1 Integration method

The integration method requires integrating the differential equations of beam deflection up to four times, depending on whether Equation 6.1, 6.2 or 6.3 is used. The general procedure of the integration method is as follows:

- Establish the equation of beam deflection in the form of Equation 6.1, 6.2 or 6.3 as appropriate. The choice of the above differential equations depends on whether or not an expression of bending moment, shear force, or applied distributed load can be formulated.
- Integrate twice, thrice and four times when Equations 6.1 to 6.3 are, respectively, used.
- Apply continuity conditions at the critical sections where shear force, bending moment or applied load changes patterns of distribution. This means that at any rigid connection of two parts of a beam the deflection and the slope of the deflection curve must be continuous. If they are joined by a pin, the deflection at the joint must be continuous.
- Apply support (boundary) conditions for the solution of the integral constants introduced in Step 2.

The most commonly seen support conditions are listed in Table 4.1. Table 6.1 lists some typical continuity conditions.

6.2.2 Superposition method

The superposition method can be used to obtain deflection of a beam subjected to multiple loads, particularly when the deflections of the beam for all or part of the individual loads are known from previous calculations or design tables and the like. For example, the beam shown in Figure 6.3a can be separated into three different cases. The algebraic sum of the three separate deflections caused by the separate loads gives the total deflection (Figure 6.3b).

Table 6.2 presents beam deflections for a number of typical load-support conditions that can be used as the separate cases to form the total solution of a complex problem.

Table 6.1 Continuity conditions at critical sections

Type of critical section		Displacements	Internal forces
Intermediate roller support		Deflection $y_a = y_b = 0$ Slope $\dfrac{dy_a}{dx} = \dfrac{dy_b}{dx}$	Bending moment $M_a = M_b$
Intermediate pin		Deflection $y_a = y_b = 0$	Shear force $V_a - V_b$ Bending moment $M_a = M_b = 0$
Concentrated force		Deflection $y_a = y_b$ Slope $\dfrac{dy_a}{dx} = \dfrac{dy_b}{dx}$	Shear force $V_a - V_b = P$ Bending moment $M_a = M_b$
Concentrated moment		Deflection $y_a = y_b$ Slope $\dfrac{dy_a}{dx} = \dfrac{dy_b}{dx}$	Shear force $V_a = V_b$ Bending moment $M_a - M_b = M$
Discontinuity in distributed load		Deflection $y_a = y_b$ Slope $\dfrac{dy_a}{dx} = \dfrac{dy_b}{dx}$	Shear force $V_a = V_b$ Bending moment $M_a = M_b$

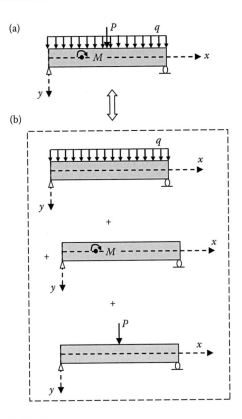

Figure 6.3 Liner superposition of deflection.

The general procedure of the superposition method is as follows:

- Resolve a complex problem into several simpler problems whose deflections are readily available.
- Express the deflections of the simpler problems in a common coordinate system.
- Superimpose the deflections algebraically to compute the total deflection.

6.2.3 Macaulay's method (step function method)

A step function is denoted by the brackets '< >'. For a variable $x - a$, we define

$$\langle x - a \rangle^n = \begin{cases} (x - a)^n & x \geq a \\ 0 & x < a \end{cases} \tag{6.4}$$

where n is any real number. This function enables us to write the bending moment for the beam shown in Figure 6.4 in a single equation.

Table 6.2 Deflection of beam

Load and support	Deflection	Maximum deflection	End rotation
	$y = \dfrac{Px^2}{6EI}(3a - x),\quad 0 \le x \le a$ $y = \dfrac{Pa^2}{6EI}(3x - a),\quad a \le x \le l$	$y_{max} = y_B$ $= \dfrac{Fa^2}{6EI}(3l - a)$	$\theta_B = \dfrac{Fa^2}{2EI}$
	$y = \dfrac{qx^2}{24EI}(x^2 + 6l^2 - 4lx)$	$y_{max} = y_B$ $= \dfrac{ql^4}{8EI}$	$\theta_B = \dfrac{ql^3}{6EI}$
	$y = \dfrac{Pbx}{6lEI}(l^2 - b^2 - x^2)\qquad \le x \le a$ $y = \dfrac{P}{EI}\left[\dfrac{bx}{6l}(l^2 - b^2 - x^2) + \dfrac{1}{6}(x - a)^3\right]$ $a \le x \le l$	$y_{max} = \dfrac{Pb}{9\sqrt{3}EIl} \times \sqrt{(l^2 - b^2)^3}$ At $\quad x = \sqrt{\dfrac{l^2 - b^2}{3}}$	$\theta_A = \dfrac{Pab(l + b)}{6lEI}$ $\theta_B = \dfrac{-Pab(l + a)}{6lEI}$
	$y = \dfrac{qx}{24EI}(l^3 - 2lx^2 + x^3)$	$y_{max} = \dfrac{5ql^4}{384EI}$	$\theta_A = -\theta_B$ $= \dfrac{ql^3}{24EI}$

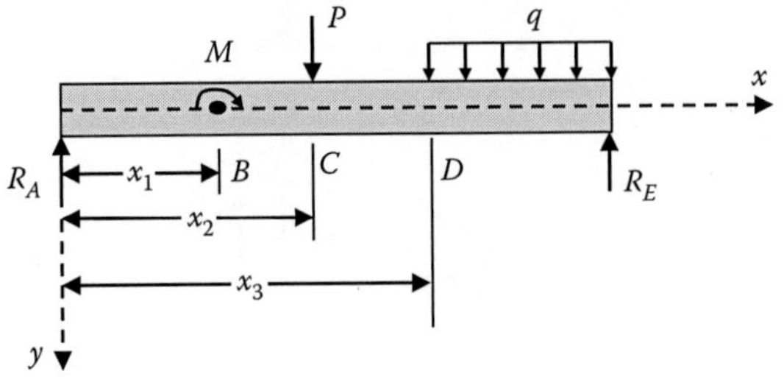

Figure 6.4 Coordinates for Macaulay's equation.

The bending moments within different parts of the beam are as follows:

- Moment on sections within (A, B) $0 \leq x \leq x_1$ $R_A x$
- Moment on sections within (B, C) $x_2 \leq x \leq x_2$ $R_A x + M$
- Moment on sections within (C, D) $x_2 \leq x \leq x_3$ $R_A x + M - P(x - x_2)$
- Moment on sections within (D, E) $x \geq x_3$ $R_A x + M - P(x - x_2) - (1/2)q(x - x_3)^2$

By using the step function described in Equation 6.4, the bending moment on an arbitrary cross section between A and E, which is x away from the left end of the beam, is

$$M(x) = R_A x + M\langle x - x_1\rangle^0 - P\langle x - x_2\rangle - \frac{q}{2}\langle x - x_3\rangle^2 \qquad (6.5)$$

From Equation 6.1, thus

$$\frac{d^2y(x)}{dx^2} = -\frac{1}{EI}\left[R_A x + M\langle x - x_1\rangle^0 - P\langle x - x_2\rangle - \frac{q}{2}\langle x - x_3\rangle^2\right]$$

$$\frac{dy(x)}{dx} = -\frac{1}{EI}\left[\frac{R_A}{2}x^2 + M\langle x - x_1\rangle^1 - \frac{P}{2}\langle x - x_2\rangle^2 - \frac{q}{6}\langle x - x_3\rangle^3\right] + C_1$$

$$y(x) = -\frac{1}{EI}\left[\frac{R_A}{6}x^3 + \frac{M}{2}\langle x - x_1\rangle^2 - \frac{P}{6}\langle x - x_2\rangle^3 - \frac{q}{24}\langle x - x_3\rangle^4\right] + C_1 x + C_2$$

The unknown constants C_1 and C_2 are determined using the support conditions.
The general procedure of Macaulay's method is as follows:

- Set up a coordinate system as shown in Figure 6.4.
- Calculate the support reactions.
- Express bending moment in a single expression in terms of the step function.
- Integrate the equation of deflection (Equation 6.1).
- Determine the two integration constants from support conditions.
- Insert the obtained integration constants back into the solution of deflection.

6.3 KEY POINTS REVIEW

- Under the action of bending moment, the axis of a beam deflects to a smooth and continuous curve.
- The perpendicular displacement of a beam axis away from its original position is called *deflection*.
- Due to deflection, cross sections of a beam rotate about its neutral axis. The rotation is called *slope,* which approximately equals the first derivative of deflection with respect to the coordinate in the axial direction.
- Curvature of a deflected beam is proportional to the bending moment acting on its cross section, while inversely proportional to the flexural rigidity of the beam, *EI*.
- Bending to an arc of a circle occurs when *M/EI* is a constant (curvature is a constant).
- Deflection of a beam depends on not only *M/EI*, but also support conditions.
- Deflection of a beam can be reduced by either using a stiffer section or adding intermediate supports.
- Deflection of a beam subjected to multiple loads is a summation of the deflections of the same beam subjected to each of the loads individually. (Superposition does not apply in inelastic or nonlinear problems.)
- The equation of deflection (Equation 6.1) is only applicable to deflection due to bending. The deflection due to shear forces is, however, comparatively small in most practical cases.

6.4 EXAMPLES

6.4.1 Examples of the integration method

EXAMPLE 6.1

Derive expressions for the slope and deflection of a uniform cantilever of length *L*, loaded with a uniformly distributed load (UDL). The flexural rigidity of the beam is *EI*. Calculate also the slope and deflection at the free end.

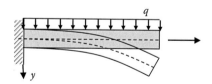

Solution

This is a simple problem. Integration can be performed on either Equation 6.1 or 6.3. Starting from Equation 6.1, bending moment along the *x*-axis must be sought first and two support conditions are needed to complete the solution. If Equation 6.3 is used, four support conditions are needed.

1. Expressing bending moment in terms of *x* and integrating Equation 6.1
   (Two boundary conditions are needed. They can be at $x = 0$, $y = 0$ and at $x = 0$, $dy/dx = 0$.)

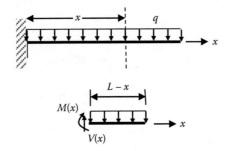

Taking the origin of the coordinate system at the fixed end, the bending moment at an arbitrary section, distance x from the origin, is

$$M(x) = -q(L - x) \times \frac{L - x}{2} = -\frac{q(L - x)^2}{2}$$

Substituting for $M(x)$ in Equation 6.1

$$\frac{d^2y}{dx^2} = -\frac{M(x)}{EI} = \frac{q}{2EI}(L^2 - 2Lx + x^2)$$

Since EI is a constant, by direct integration

$$\frac{dy}{dx} = \frac{q}{2EI}\int (L^2 - 2Lx + x^2)\,dx = \frac{q}{2EI}\left(L^2x - Lx^2 + \frac{x^3}{3}\right) + C_1$$

where C_1 is a constant of integration and can be determined using the condition that the slope is zero at the fixed end, that is, at the origin.

Since $dy/dx = 0$ at $x = 0$

$$C_1 = 0$$

Thus,

$$\frac{dy}{dx} = \frac{q}{2EI}\left(L^2x - Lx^2 + \frac{x^3}{3}\right)$$

Integrating again

$$y = \frac{q}{2EI}\int \left(L^2x - Lx^2 + \frac{x^3}{3}\right)dx = \frac{q}{2EI}\left(\frac{L^2}{2}x^2 - \frac{L}{3}x^3 + \frac{1}{12}x^4\right) + C_2$$

Here C_2 is the second constant of integration, which can be determined using the condition that the deflection is zero at the origin.

Since $y = 0$ at $x = 0$

$$C_2 = 0$$

Thus, the deflection of the beam is

$$y = \frac{q}{2EI}\left(\frac{L^2}{2}x^2 - \frac{L}{3}x^3 + \frac{1}{12}x^4\right)$$

2. Integration of Equation 6.3 (Four boundary conditions are needed. They can be at $x = 0$, $y = dy/dx = 0$ and at $x = L$, $M = V = 0$.)

$$EI\frac{d^4y}{dx^4} = q$$

Since EI is a constant, by direct integration and using Equation 6.2

$$EI\frac{d^3y}{dx^3} = \int q\,dx = qx + C_1 = -V(x)$$

The constant C_1 is determined by using the condition that the shear force is zero at the free end.

Since $V = 0$ at $x = L$

$$qL + C_1 = -V(L) = 0$$

Thus, $C_1 = -qL$.
   Then,

$$EI\frac{d^3y}{dx^3} = qx - qL$$

Integrating the second time and considering Equation 6.1

$$EI\frac{d^2y}{dx^2} = \frac{q}{2}x^2 - qLx + C_2 = -M(x)$$

Since $M = 0$ at $x = L$

$$\frac{q}{2}L^2 - qLL + C_2 = -M(L) = 0$$

Thus, $C_2 = qL^2/2$.
   Then,

$$EI\frac{d^2y}{dx^2} = \frac{q}{2}x^2 - qLx + \frac{qL^2}{2}$$

Integrating the third time

$$EI\frac{dy}{dx} = \frac{q}{6}x^3 - \frac{qL}{2}x^2 + \frac{qL^2}{2}x + C_3$$

Since $dy/dx = 0$ at $x = 0$

$$C_3 = 0$$

Thus, the expression of slope for the beam is

$$EI\frac{dy}{dx} = \frac{q}{6}x^3 - \frac{qL}{2}x^2 + \frac{qL^2}{2}x$$

or

$$\frac{dy}{dx} = \frac{q}{2EI}\left(\frac{1}{3}x^3 - Lx^2 + L^2x\right)$$

Finally,

$$EIy = \frac{q}{24}x^4 - \frac{qL}{6}x^3 + \frac{qL^2}{4}x^2 + C_4$$

Since $y = 0$ at $x = 0$

$$C_4 = 0$$

The deflection of the beam is therefore

$$y = \frac{1}{EI}\left(\frac{q}{24}x^4 - \frac{qL}{6}x^3 + \frac{qL^2}{4}x^2\right)$$

$$= \frac{q}{2EI}\left(\frac{1}{12}x^4 - \frac{L}{3}x^3 + \frac{L^2}{2}x^2\right)$$

which is the same as that obtained from a.
At the free end $x = L$, we have
Slope at the free end

$$\theta_L = \left.\frac{dy}{dx}\right|_{x=L} = \frac{q}{2EI}\left(L^2L - LL^2 + \frac{L^3}{3}\right) = \frac{qL^3}{6EI}$$

Deflection at the free end

$$y|_{x=L} = \frac{q}{2EI}\left(\frac{L^2}{2}L^2 - \frac{L}{3}L^3 + \frac{1}{12}L^4\right) = \frac{qL^4}{8EI}$$

The slope and deflection are identical to those shown in Table 6.2.

EXAMPLE 6.2

A simply supported beam is loaded with a concentrated force as shown in figure below. The flexural rigidity EI is constant. Find the expression of deflection of the beam.

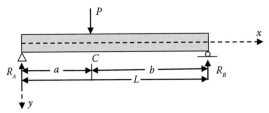

Solution

The concentrated load applied along the beam results in different expressions of bending moment for the segments between AC and CB. The integration of the equation of deflection must proceed for within AC and CB, which generates four unknown constants of integration. The four conditions that can be used to determine the constants are $y(x = 0) = 0$, $y(x = L) = 0$, and the continuity of deflection and slope at C.

Consider the entire beam and take moment about the right-hand-side end, which yields

$$R_A L - P_b = 0$$

$$R_A = \frac{Pb}{L}$$

For a cross section within AC, the bending moment is

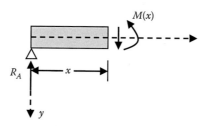

$$M(x) = R_A x = \frac{Pb}{L} x$$

From Equation 6.1

$$EI \frac{d^2 y}{dx^2} = -\frac{Pb}{L} x$$

So,

$$EI \frac{dy}{dx} = -\frac{Pb}{2L} x^2 + C_1 \tag{6.6}$$

$$EIy = -\frac{Pb}{6L} x^3 + C_1 x + C_2 \tag{6.7}$$

Since $y = 0$ at $x = 0$

$$C_2 = 0$$

Thus, at $x = a$

$$EIy \big|_{x=a} = -\frac{Pb}{6L} a^3 + C_1 a$$

$$EI \frac{dy}{dx} \bigg|_{x=a} = -\frac{Pb}{2L} a^2 + C_1$$

For a cross section within CB, the bending moment is

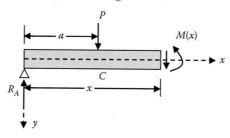

$$M(x) = R_A x - P(x - a) = \frac{Pb}{L}x - P(x - a)$$

$$= Pa - \frac{Pa}{L}x$$

From Equation 6.1

$$EI\frac{d^2y}{dx^2} = -Pa + \frac{Pa}{L}x$$

So,

$$EI\frac{dy}{dx} = -Pax + \frac{Pa}{2L}x^2 + C_3$$

$$EIy = -\frac{Pa}{2}x^2 + \frac{Pa}{6L}x^3 + C_3x + C_4$$

Since $y = 0$ at $x = L$

$$-\frac{PaL^2}{3} + C_3L + C_4 = 0 \tag{6.8}$$

At $x = a$

$$EI\frac{dy}{dx}\bigg|_{x=a} = -Pa^2 + \frac{Pa}{2L}a^2 + C_3 \tag{6.9}$$

$$EIy\big|_{x=a} = -\frac{Pa}{2}a^2 + \frac{Pa}{6L}a^3 + C_3a + C_4 \tag{6.10}$$

Equating deflections of both AC and CB at $x = a$ yields (Equations 6.7 and 6.10)

$$-\frac{Pb}{6L}a^3 + C_1a = -\frac{Pa}{2}a^2 + \frac{Pa}{6L}a^3 + C_3a + C_4 \tag{6.11}$$

Equating slopes of both AC and CB at $x = a$ yields (Equations 6.6 and 6.9)

$$-\frac{Pb}{2L}a^2 + C_1 = -Pa^2 + \frac{Pa}{2L}a^2 + C_3 \tag{6.12}$$

The solution of Equations 6.8, 6.11 and 6.12 gives

$$C_1 = \frac{Pb}{6L}(L^2 - b^2)$$

$$C_3 = \frac{Pa}{6L}(2L^2 + a^2)$$

$$C_4 = \frac{Pa^3}{6}$$

With these constants, the expressions of deflection of the beam are

$$y = \frac{Pbx}{6EIL}[(L^2 - b^2) - x^2] \quad 0 \le x \le a$$

$$y = \frac{Pb}{6EIL}\left[\frac{L}{b}(x-a)^3 + (L^2 - b^2)x - x^3\right] \quad a \le x \le L$$

It can be concluded from the above example that when applied loads have abrupt changes, including concentrated forces, moments and patched distributed loads, along a beam, using the direct integration method normally involves solving simultaneous linear algebraic equations. To avoid this, Macaulay's method is a better option.

6.4.2 Examples of the superposition method

The superposition method is quite useful when a complex load can be resolved into a superposition of several simple loads, and the deflections due to these simple loads are known. This method is particularly useful for calculating deflections and slopes at given points along a beam.

EXAMPLE 6.3

A uniform cantilever is subjected to a uniformly distributed pressure and a concentrated force applied at the free end (figure below). Calculate the deflection and slope of the beam at the free end.

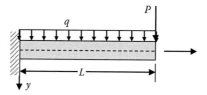

Solution

This is a simple question showing the basic principle of superposition. The beam can be assumed as under a combined action of the uniformly distributed pressure and the point load applied at the free end. The deflection and slope of the beam under a single action of either the pressure or the point load can be found from Table 6.2.

The problem can be resolved into two simple problems as

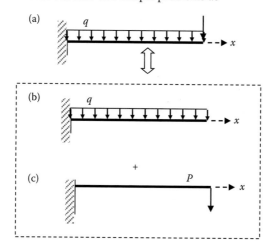

From Table 6.2, the free end deflection and rotation due to the uniform pressure are, respectively,

$$y_{(a)} = \frac{qL^4}{8EI}$$

$$\theta_{(a)} = \frac{qL^3}{6EI}$$

The deflection and slope at the free end due to the concentrated force can also be found from Table 6.2 (let $b = 0$ and $a = L$ in case 1), which are

$$y_{(b)} = \frac{PL^3}{3EI}$$

$$\theta_{(b)} = \frac{PL^2}{2EI}$$

Thus, the total deflection and slope of the beam at the free end are, respectively,

$$y_{\text{Total}} = y_{(a)} + y_{(b)} = \frac{qL^4}{8EI} + \frac{PL^3}{3EI}$$

$$\theta_{\text{Total}} = \theta_{(a)} + \theta_{(b)} = \frac{qL^3}{6EI} + \frac{PL^2}{2EI}$$

EXAMPLE 6.4

Calculate the free end deflection of the beam shown in figure below. The beam has an I uniform flexural rigidity.

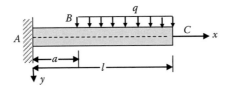

Solution

To use the solution presented in Table 6.2, the UDL is extended to the fixed end (case [a]) and is effectively removed by applying an upward UDL of equal magnitude between A and B (case [b]).

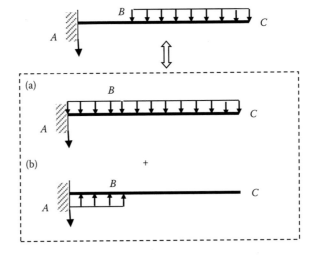

From Table 6.2, the deflection and the slope at the free end of case (a) are, respectively,

$$y_C^{(a)} = \frac{ql^4}{8EI}$$

$$\theta_C^{(a)} = \frac{ql^3}{6EI}$$

For case (b) since the segment between B and C is not loaded and the right-hand-side end is completely free, the slope at C is identical to the slope at B. The deflection at the free end equals the deflection at B plus the product of the slope at B and the length of BC:

$$\theta_C^{(b)} = \theta_C^{(b)} = -\frac{qa^3}{6EI}$$

$$y_C^{(b)} = y_B^{(b)} + \theta_B^{(b)}(l - a) = -\frac{qa^4}{8EI} - \frac{qa^3}{6EI}(l - a)$$

The total deflection and slope at the free end are, respectively,

$$y_C = y_C^{(a)} + y_C^{(b)} = \frac{ql^4}{8EI} - \frac{qa^4}{8EI} - \frac{qa^3}{6EI}(l - a) = \frac{q}{24EI}(3l^4 - 4la^3 + a^4)$$

$$\theta_C = \theta_C^{(a)} + \theta_C^{(b)} = \frac{ql^3}{6EI} - \frac{qa^3}{6EI} = \frac{q}{6EI}(l^3 - a^3)$$

EXAMPLE 6.5

A simply supported uniform beam is subjected to a uniform pressure from the mid-span to the right-hand-side support as shown in figure below. Use the superposition method to calculate the mid-span deflection.

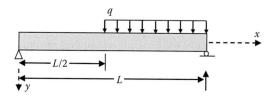

Solution

There are no solutions in Table 6.2 that can be used directly. However, the UDL can be taken as a sum of an infinite number of small concentrated forces acting on the beam. The deflections at the mid-span due to these concentrated forces can be found from Table 6.2. The summation of the infinite number of deflections can be calculated in the form of integration.

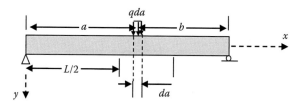

The beam shown above is subjected to a concentrated force, qda, at a distance a away from the left-hand-side support. From Table 6.2 (case 3), the following mid-span deflection can be obtained:

$$dy_m = \frac{qda(L-a)L/2}{6LEI}\left[L^2 - (L-a)^2 - \frac{L^2}{4}\right]$$

$$= \frac{q}{EI}\left[\frac{L^2}{16}(L-a) - \frac{1}{12}(L-a)^3\right]da$$

when P is replaced by qda, l by L, x by $L/2$ and b by $L-a$. Thus, the total mid-span deflection due to the distributed load is

$$y_m = \int_{L/2}^{L} \frac{q}{EI}\left[\frac{L^2}{16}(L-a) - \frac{1}{12}(L-a)^3\right]da = \frac{5qL^4}{768EI}$$

EXAMPLE 6.6

Find the deflection for the uniformly loaded, two-span continuous beam shown in figure below. EI is constant.

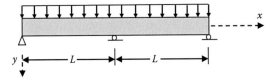

Solution

This is a statically indeterminate beam whose reactions cannot be determined using the equilibrium conditions. Superposition method can, sometimes, be used conveniently to calculate the reactions. The beam shown in figure below can be taken as the superposition of a beam subjected to the uniform downward pressure and the same beam subjected to an unknown concentrated upward force at the mid-span. The combined action of the two loads results in a zero deflection at the mid-span of the beam.

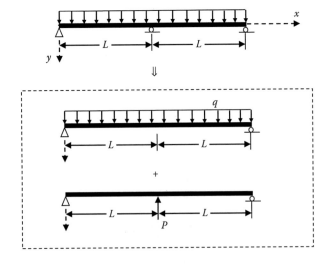

From Table 6.2, the mid-span deflections of the beam due to the UDL and the concentrated force are, respectively,

$$y_{UDL} = \frac{5q(2L)^4}{384EI} = \frac{5qL^4}{24EI}$$

$$y_P = -\frac{p(2L)^3}{48EI} = -\frac{PL^3}{6EI}$$

The sum of these two deflections must be zero because the beam is roller-pinned at the mid-span. So,

$$\frac{5qL^4}{24EI} - \frac{PL^3}{6EI} = 0$$

$$P = \frac{5qL}{4}$$

The deflection of the beam is equal to the deflection due to the uniform load and the deflection of the beam subjected to an upward concentrated force, $5qL/4$, at the mid-span. From Table 6.2, for example, the deflection of the left-hand-side span is

$$y_{UDL}(x) = \frac{qx}{24EI}(l^3 - 2lx^2 + x^3) = \frac{qx}{24EI}[(2L)^3 - 2(2L)x^2 + x^3]$$

$$= \frac{qx}{24EI}(8L^3 - 4Lx^2 + x^3)$$

$$y_P(x) = \frac{Pbx}{6lEI}(l^2 - b^2 - x^2) = \frac{PLx}{6(2L)EI}[(2L)^2 - L^2 - x^2]$$

$$= \frac{5qx}{48EI}[3L^3 - x^2L]$$

$$y(x) = y_{UDL}(x) - y_P(x) = \frac{q}{48}(2x^4 - 3Lx^3 + L^3x) \quad 0 \le x \le L$$

Due to symmetry, the deflection of the right-hand-side span is numerically identical to the above though the expression of the curve is different ($y_P(x)$ for $L \le x \le 2L$ should be used in the above superposition).

6.4.3 Examples of Macaulay's method

EXAMPLE 6.7

A uniform beam 16 ft long is simply supported at its ends and carries a uniform distributed load of $q = 0.5$ ton/ft between B and C, which are 3 and 11 ft from A, respectively. The beam also carries a concentrated force of $P = 6$ tons 13 ft from A. If $E = 13,400$ tons/in.^{2} and $I = 204.8$ in.^{4}, obtain the deflection of the beam and calculate the mid-span deflection.

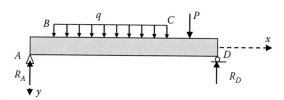

Solution

The change of load patterns can be dealt with by the step functions (Equation 6.4) and the general expression of bending moment (Equation 6.5). When using Macaulay's method with distributed loads, it is essential that the distributed load is continued to the end of the beam. In this example, the distributed load is extended to cover the extra range CD. To remove the additional loading, an upward pressure of the same magnitude must also be applied in the same range.

The equivalent loading of the beam is shown below:

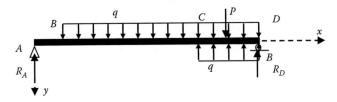

To use Equation 6.5, the reaction force at A must be obtained first.
Taking moment about D (figure above)

$$16R_A = 0.5 \times 8 \times 9 + 6 \times 3$$
$$R_A = 3.375 \text{ tons}$$

Hence from Equation 6.5, for an arbitrary section at a distance x from A

$$M(x) = R_A x - P\langle x - 13 \rangle - \frac{q}{2}\langle x - 3 \rangle^2 + \frac{q}{2}\langle x - 11 \rangle^2$$
$$= 3.375x - 6\langle x - 13 \rangle - \frac{0.5}{2}\langle x - 3 \rangle^2 + \frac{0.5}{2}\langle x - 11 \rangle^2$$

From Equation 6.1

$$EI\frac{d^2y}{dx^2} = -M(x)$$
$$= -3.375x + 6\langle x - 13 \rangle + \frac{0.5}{2}\langle x - 3 \rangle^2 - \frac{0.5}{2}\langle x - 11 \rangle^2$$

Integrating once

$$EI\frac{dy}{dx} = -\frac{3.375}{2}x^2 + 3\langle x - 13 \rangle^2 + \frac{1}{12}\langle x - 3 \rangle^3 - \frac{1}{12}\langle x - 11 \rangle^3 + C_1$$

Integrating twice

$$EIy = -\frac{3.375}{6}x^3 + \langle x - 13 \rangle^3 + \frac{1}{48}\langle x - 3 \rangle^4 - \frac{1}{48}\langle x - 11 \rangle^4 + C_1 x + C_2$$

The two constants of integration can be determined by introducing support conditions:
At $x = 0$, $y = 0$
$C_2 = 0$ (use the properties of step function defined in Equation 6.4)
At $x = 16$, $y = 0$

$$0 = -\frac{3.375}{6} \times 16^3 + \langle 16 - 13 \rangle^3 + \frac{1}{48} \langle 16 - 3 \rangle^4 - \frac{1}{48} \langle 16 - 11 \rangle^4 + 16C_1$$
$$C_1 = 105.94$$

The deflection of the beam is

$$y(x) = \frac{1}{EI} \left[-0.5625x^3 + \langle x - 13 \rangle^3 + \frac{1}{48} \langle x - 3 \rangle^4 - \frac{1}{48} \langle x - 11 \rangle^4 + 105.94x \right]$$

At the mid-span, $x = 8$ in.

$$y_{mid} = \frac{12^2}{13,400 \times 204.8} \left[-0.5625 \times 8^3 + \frac{1}{48} 5^4 + 105.94 \times 8 \right] = 0.03\,\text{ft}$$
$$= 0.361\,\text{in.}$$

EXAMPLE 6.8

A built-in beam of length L carries a concentrated load P at distance a from the left-hand-side end (figure below). Obtain the deflection of the beam and calculate the fixed end moments.

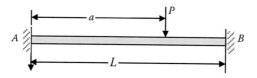

Solution

This is a statically indeterminate problem. The reaction forces at A and B cannot be solved by static equilibrium. At both ends, unknown bending moments and shear forces exist. Before these forces are found, the general expression of bending moment is expressed in terms of these unknown reactions that can be determined subsequently by introducing support conditions.

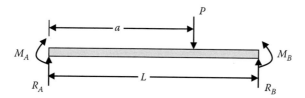

Assume that the two reaction forces at A are, respectively, R_A and M_A. From Equation 6.5, the general expression of bending moment is

$$M(x) = R_A x + M_A \langle x - 0 \rangle^0 - P\langle x - a \rangle = R_A x + M_A - P\langle x - a \rangle$$

From Equation 6.1

$$EI \frac{d^2 y}{dx^2} = -M(x) = -R_A x - M_A + P\langle x - a \rangle$$

Integrating once

$$EI\frac{dy}{dx} = -\frac{R_A}{2}x^2 - M_Ax + \frac{P}{2}\langle x - a \rangle^2 + C_1 \tag{6.13}$$

At the fixed end

$$x = 0, \quad \frac{dy}{dx} = 0$$

Thus, from Equation 6.13

$$C_1 = 0$$

Integrating twice

$$EIy = -\frac{R_A}{6}x^3 - \frac{M_A}{2}x^2 + \frac{P}{6}\langle x - a \rangle^3 + C_2 \tag{6.14}$$

Again at the fixed end

$$x = 0, y = 0$$

From Equation 6.14

$$C_2 = 0$$

The expression of deflection in terms of the unknown reaction forces at A is

$$y(x) = \frac{1}{EI}\left[-\frac{R_A}{6}x^3 - \frac{M_A}{2}x^2 + \frac{P}{6}\langle x - a \rangle^3 \right] \tag{6.15}$$

The reaction force, R_A, and the fixed end moment, M_A, can be determined through the introduction of the support conditions at B.
At $x = L$

$$\frac{dy}{dx} = 0 \quad \text{and} \quad y = 0$$

From Equation 6.13

$$0 = -\frac{R_A}{2}L^2 - M_AL + \frac{P}{2}(L - a)^2$$

From Equation 6.14

$$0 = -\frac{R_A}{6}L^3 - \frac{M_A}{2}L^2 + \frac{P}{6}(L - a)^3$$

Solving the above simultaneous equations in terms of R_A and M_A yields

$$M_A = -\frac{Pa(L - a)^2}{L^2}$$

$$R_A = \frac{P(L - a)^2}{L^3}(2a + L)$$

Once M_A and R_A are found, M_B and R_B can be easily obtained by considering the equilibrium of the entire beam, which are, respectively,

$$M_B = -\frac{P(L-a)a^2}{L^2}$$

$$R_B = \frac{Pa^2}{L^3}(3L-2a)$$

Substituting the above solutions into Equation 6.15 yields the deflection of the beam.

6.5 CONCEPTUAL QUESTIONS

1. Describe what is meant by a 'simply supported' support.
2. Describe what is meant by a 'fixed end' support.
3. What is the support condition of a free end?
4. What are deflection, slope and curvature of a beam?
5. How are deflection, slope and curvature related to each other for a beam in bending?
6. What is the difference between axial stiffness and flexural rigidity of a member?
7. What is the difference between torsional stiffness and flexural rigidity of a member?
8. Under what condition will a beam bend into a circular arc?
9. Is the curvature of a beam zero at the location where bending moment is zero?
10. Does maximum deflection always occur at the position where slope is zero?
11. Does maximum deflection always occur at the position where bending moment is maximum?
12. If the integration method is applied to Equation 6.1 to find deflection of the beam shown in the figure below, how many constants of integration in total are to be determined from imposing support and continuity conditions? And what are the conditions?

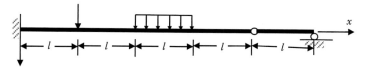

13. If two beams have the same length and flexural rigidity and are subjected to the same external loads, are the deflections of the two beams identical, and why?

6.6 MINI TEST

PROBLEM 6.1

The cantilever shown in figure below has a rectangular cross section. With all the conditions staying the same, except that the depth of the beam is doubled, complete the following statements:

1. The maximum normal stress is now _____ times the maximum normal stress in the original beams.

2. The maximum shear stress is now _____ times the maximum shear stress in the original beams.
3. The maximum deflection is now _____ times the maximum deflection of the original beam.
4. The maximum slope is now _____ times the maximum slope of the original beam.

PROBLEM 6.2

Calculate the mid-span deflection of the beams shown in figure (a, b) below using the superposition method and the solutions from Table 6.2.

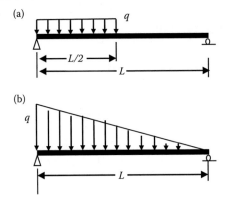

PROBLEM 6.3

Figure below shows a horizontal beam freely supported at its ends by the free ends of two cantilevers. If the flexural rigidity, EI, of the cantilevers is twice that of the beam, calculate the deflection at the centre of the beam.

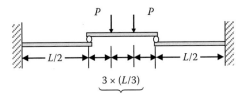

PROBLEM 6.4

Determine the deflection along the beam, and the magnitude of deflection at D of the beam shown in figure below. The flexural rigidity (EI) is 1.0 MN m^2.

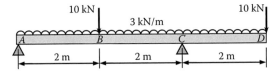

PROBLEM 6.5

Determine the deflection curve of the statically indeterminate beam shown in figure below. The beam has a constant EI.

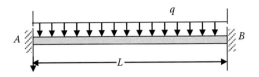

Chapter 7

Complex stresses

In a practical design, a structure is usually subjected to a combination of different types of loading that generate various types of stresses from different directions within the structure. For example, the stress field in a beam–column joint is very complex, with a combination of bending, shearing and contact stresses (Figure 7.1a). If a cut (plane) is taken through a point, the stress on the plane is usually different from the stress on a different plane through the same point, not just in terms of magnitude but also direction. On an arbitrary plane through a point, a general stress (σ) can always be resolved into three independent components that are perpendicular to each other (Figure 7.1b). The three components include a normal stress (σ_n), which is perpendicular to the plane, two shear stresses (τ_1 and τ_2), which are parallel to the plane and perpendicular to each other. The stresses at a point inside the joint are best presented by the stresses acting on an infinitesimal cubic element taken around the point. The element has six faces (planes) that are either perpendicular or parallel to each other. On each of the faces there are three independent stresses, including two shear stresses and a normal stress.

Figure 7.2 shows all the stresses at a point in a material, which is sufficient and necessary to represent the *state of stress* at the point.

- When a structure is subjected to external loads, the state of stress is, in general, different at different points within the structure.
- At a point in the structure, the stress in one direction is usually different from the stress in a different direction.
- State of stress (Figure 7.1c) shows stresses acting on six different planes at a point. Therefore, when we say that we know a stress, it means that we know not only the magnitude and direction of the stress, but also the plane on which the stress acts.
- Since the cubic element has infinitesimal dimensions in the three coordinate directions, the normal stresses acting on any two faces that are parallel to each other are equal but in opposite directions. On any two planes that are perpendicular to each other, the shear stresses perpendicular to the intersection of the two planes are equal, but in an opposite sense, that is, are either towards or away from the intersection line. At a point, therefore, there are only six independent stresses, that is, σ_x, σ_y, σ_z, τ_{xy}, $(= \tau_{yx})$, $\tau_{xz}(= \tau_{zx})$ and τ_{xz}, $(= \tau_{zy})$ (Figure 7.2).
- At any point within a material, if the two shear stresses are zero on a plane, this plane is called *principal plane*. The normal stress acting on the principal plane is called *principal stress*, and its direction is called *principal direction*. If a cubic element is chosen such that all the faces of the cube are free of shear stresses, the element is called *principal element*.

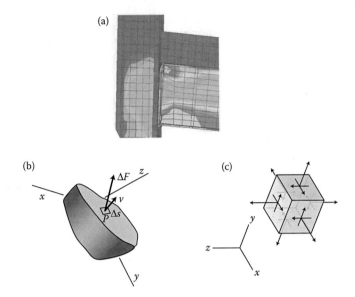

Figure 7.1 Three-dimensional stresses at a point of material.

7.1 TWO-DIMENSIONAL STATE OF STRESS

In some cases, the stresses relative to a particular direction are sufficiently small compared to the stresses relative to the other two directions. Typical example problems include stresses in a thin plate subjected to in-plane loadings (Figure 7.3a) and in a thin-walled vessel under internal pressure (Figure 7.3b) or torsion (Figure 7.3c). Suppose that the small stress is related to the z direction and is ignored, the three-dimensional state of stress can be reduced to a two-dimensional one. Since the remaining stresses lie in a plane, the simplified two-dimensional problem is called *plane problem*. For the thin plate subjected to in-plane loads, Figure 7.3a shows the two-dimensional state of stress. For a thin-walled cylinder subjected to internal pressure or torsion, the states of stress are shown by Figure 7.3b, c.

In Figure 7.3, the normal stresses (σ_x and σ_y) have a single subscript index that indicates the coordinate axis the stresses are parallel to. The first subscript index of a shear stress (τ_{xy} or τ_{yx}) denotes the direction of the normal of the plane on which the stress acts, while the second index denotes the axis to which the shear stress is parallel. Since the two-dimensional element is infinitesimal, τ_{xy} is numerically equal to τ_{yx}.

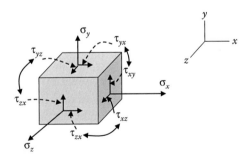

Figure 7.2 Three-dimensional state of stresses.

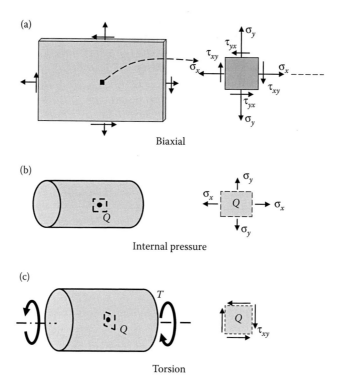

Figure 7.3 Examples of two-dimensional state of stresses.

7.1.1 Sign convention of stresses

The stresses shown in Figure 7.3 are all defined as positive in the chosen coordinate system, where the following sign conventions are followed:

- Tensile and compressive stresses are always defined, respectively, as positive and negative.

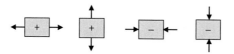

- Positive shear stresses are defined similar to the positive shear forces defined in Section 4.4.

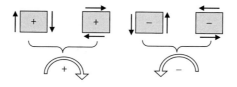

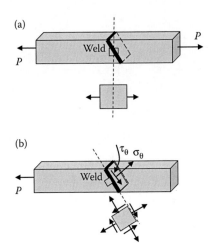

Figure 7.4 Example of stresses on an inclined plane.

7.1.2 Analytical method

Once the stress components that align with a typical x–y coordinate system are found (Figure 7.2), transforming the stresses into another coordinate system is sometimes necessary. Two key reasons that we may want to calculate stresses in a different coordinate system include the following:

> - To determine the stress in an important direction, for example, stresses normal and parallel to the plane of a weld (Figure 7.4b).
> - To determine the maximum normal stress or maximum shear stress at a point. These stresses may not necessarily align with the chosen coordinate directions.

1. Stresses on an arbitrarily inclined plane: To further understand the stresses on an arbitrarily inclined plane (cut) through a point in a material, consider the two elements taken around the point in Figure 7.4.

   Since the elements are taken at the same point, we might take them for the same state of stress, but measured in a different coordinate system. The new coordinate system, x'–y', is defined by a rotation θ of both the coordinate axes from their original

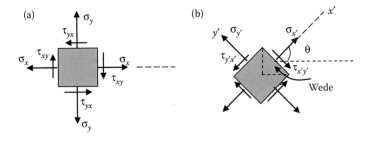

Figure 7.5 Stresses on an arbitrarily inclined plane.

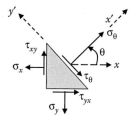

Figure 7.6 Equilibrium of a wedge.

directions (Figure 7.5). The relationship between the states of stress in terms of the original system and the rotated system can be best presented by considering the equilibrium of the wedge of material (Figure 7.6) taken from Figure 7.5b. The normal stress, σ_θ, and shear stress, τ_θ, acting on the inclined plane are, respectively, the normal stress, $\sigma_{x'}$, and shear stress, $\tau_{x'y'}$ in the rotated coordinate system, while the stresses acting on the vertical and horizontal sides of the wedge are identical to those acting on the vertical and horizontal sides of the element (Figure 7.5a) in the original coordinate system.

The equilibrium of the wedge yields

$$\sigma_\theta = \sigma_x \cos^2 \theta + \sigma_y \sin^2 \theta - \tau_{xy} \sin 2\theta$$
$$= \frac{\sigma_x + \sigma_y}{2} + \frac{\sigma_x - \sigma_y}{2} \cos 2\theta - \tau_{xy} \sin 2\theta \qquad (7.1)$$
$$\tau_\theta = \left(\frac{\sigma_x + \sigma_y}{2} \right) \sin 2\theta + \tau_{xy} \cos 2\theta$$

where an anticlockwise angle from the x-axis is defined as positive.

2. Principal stresses: From Equation 7.1, the stresses on an inclined plane change as the value of θ changes. It means that on different planes taken by cutting through the point the stresses are generally different. It is natural to think that there are special planes on which the normal stress reaches either maximum or minimum (maximum compressive stress) algebraically. The maximum and minimum normal stresses are both called *principal stresses*. When a normal stress is either maximum or minimum, the plane on which the stress acts is always free of shear stress. In a two-dimensional stress system, there are two principal stresses, that is, the maximum and the minimum normal stresses at a point, as shown in Figure 7.7.

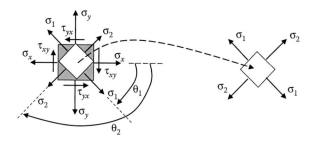

Figure 7.7 Principal stresses and directions.

The principal stresses can be calculated as follows:

$$\sigma_{max} = \sigma_1$$
$$= \frac{1}{2}\left[(\sigma_x + \sigma_y) + \sqrt{(\sigma_x - \sigma_y)^2 + 4\tau_{xy}^2}\right]$$
$$\sigma_{min} = \sigma_2$$
$$= \frac{1}{2}\left[(\sigma_x + \sigma_y) - \sqrt{(\sigma_x - \sigma_y)^2 + 4\tau_{xy}^2}\right]$$

(7.2)

3. The directions of principal stresses: The angle between a principal stress and the x-axis can be calculated as follows:

$$\tan 2\theta = \frac{-2\tau_{xy}}{\sigma_x - \sigma_y}$$
$$\theta = \frac{1}{2}\tan^{-1}\left(\frac{-2\tau_{xy}}{\sigma_x - \sigma_y}\right)$$

(7.3)

Since the two principal stresses are perpendicular to each other, the direction of the second principal stress is $\theta + 90°$.

In a plane problem, there are two principal stresses (Equation 7.2) and two associated directions. The easiest way to relate the stresses to their respective directions is based on the following simple observation.

In Figure 7.8, the shear stress τ_{xy} generates tension in one diagonal direction and compression in the other, which suggests that combined with actions of σ_x and σ_y, the normal stress in the direction of the diagonal in tension is '*more tensile*' or larger than that in the direction of the diagonal in compression. Hence, it can be concluded that the direction of σ_1 is in the quarter where the shear stresses are pointing to.

At a point in a material, a normal stress is a principal stress if

- The stress is either the maximum tensile stress or the maximum compressive stress at the point.

or

- The plane on which the normal stress acts is free of shear stresses. The plane is one of the principal planes.

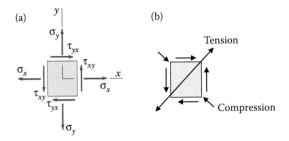

Figure 7.8 Directions of principal stresses.

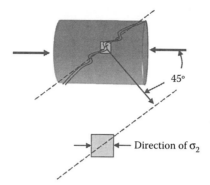

Figure 7.9 Concrete failure due to maximum shearing.

4. Maximum shear stress: Following the same argument as for the existence of maximum normal stresses, there exist special planes on which shear stress reaches maximum or minimum. (They have equal magnitudes in an opposite sense.) Figure 7.9 shows a concrete cylinder under compression. The cylinder fails due to maximum shearing at about 45° to the axial of compression. The cylinder may fail along the other diagonal direction under the same compression due to an equal shear stress of opposite sense. By observation, the plane perpendicular to the axial direction is a principal plane since there is no shear stress acting on this plane. The maximum shear stress acts on the plane that is 45° away from the principal plane. From Equation 7.1, the maximum shear stress can be obtained by replacing σ_x and σ_y with σ_1 and σ_2, respectively, and calculating the shear stress at the 45° plane

$$\tau_{max} = \frac{\sigma_1 - \sigma_2}{2} \tag{7.4}$$

Introducing Equation 7.2 into Equation 7.4:

$$\tau_{max} = \frac{1}{2}\sqrt{(\sigma_x - \sigma_y)^2 + 4\tau_{xy}^2} \tag{7.5}$$

7.1.3 Graphic method

Mohr's circle illustrates principal stresses and stress transformations via a graphical format, that is, a graphic representation of Equations 7.1 to 7.5. The circle is plotted in a plane coordinate system where the horizontal axis denotes normal stress. The vertical coordinate denotes the shear stress on the same plane (Figure 7.10). While plotting a Mohr's circle a sign convention, for example, the one defined in Section 7.1.1, must be followed. Here we take tensile stresses and shear stresses that would turn an element clockwise as positive. Figure 7.10 shows how the stresses acting on an element are related to a Mohr's circle.

The two principal stresses are shown by the σ coordinates of the two intersections of the circle with the horizontal axis (where shear stresses are zero). The vertical coordinates of either the highest or the lowest point on the circle denote the maximum magnitude of shear stress that is also equal to the radius of the circle. If the state of stress at a point is

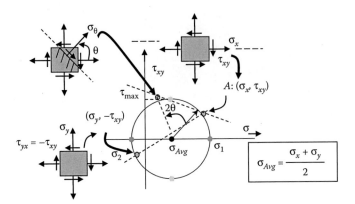

Figure 7.10 Mohr's circle.

known, that is, σ_x, σ_y and τ_{xy} are known, the following steps can be followed to plot a Mohr's circle:

- Set up a coordinate system where the horizontal axis is the normal stress axis and the vertical axis is the shear stress axis; positive directions of the axes take upwards and to the right.
- Locate two points, A and B, related to, respectively, the stresses on the right and upper faces of a state-of-stress element, with respective coordinates (σ_x, τ_{xy}) and $(\sigma_x, -\tau_{xy})$, in the σ–τ coordinate system and connect the two points by a straight line. The sign convention defined in Section 7.1.1 must be followed to locate the two points. The intersection of the straight line with the σ-axis is marked 'O' and is at a distance of $\sigma_{Avg} = (\sigma_x + \sigma_y)/2$ away from the origin.
- With its centre at 'O', draw a circle passing through points A and B.
- Measure σ coordinates of the two intersections of the circle with the σ-axis to obtain the two principal stresses.
- Measure the radius of the circle to obtain the maximum shear stress.
- To determine the magnitudes of the stresses acting on an inclined plane, $\theta°$ away from the right-hand-side face, measure an angle of $2\theta°$ from OA and take the coordinates of the intersection with the circle.
- The horizontal and vertical coordinates of the intersection are, respectively, the normal and shear stresses on the inclined plane.

7.2 KEY POINTS REVIEW

7.2.1 Complex stress system

- At a point in a material, there are six independent stress components, including three normal stresses and three shear stresses.
- In a two-dimensional case, there are three independent stresses, two normal stresses and one shear stress, at a point of the material.
- A stress usually varies from point to point.

- A stress is uniquely defined by the following three properties:
  - Magnitude
  - Direction
  - Plane (cross section/cut) on which the stress acts
- Without knowing any of the three, the stress is not completely defined.
- Principal stresses are normal stresses and include both the maximum and minimum stresses.
- In a three-dimensional stress system, there are three principal stresses; while in a two-dimensional system, there are two principal stresses.
- Principal stresses are always perpendicular to each other.
- The plane on which a principal stress acts is free of shear stresses.
- Maximum shear stress is equal to half of the maximum difference between principal stresses.
- The plane of maximum shear stress is always 45° away from a principal plane.

7.2.2 Mohr's circle

- The largest and smallest horizontal coordinates of the circle are, respectively, the two principal stresses σ_1 and σ_2.
- The maximum shear stress is numerically equal to the length of the radius of the circle and also equal to $(\sigma_1 - \sigma_2)/2$.
- An angle difference θ between two planes through a point is represented by a difference of 2θ between the two locations relative to the stresses on the two planes along Mohr's circle.
- A normal stress equal to $(\sigma_x + \sigma_y)/2$ acts on the planes of maximum shear stresses.
- If $\sigma_1 = \sigma_2$, Mohr's circle degenerates into a point, for which no shear stresses develop at the point.

7.3 EXAMPLES

EXAMPLE 7.1

At a point in a masonry structure, the stress system caused by the applied loadings is shown in figure (a) below. Calculate the magnitudes and orientations of the principal stresses at the point. If the stone, of which the structure is made, is stratified and weak in shear along the planes parallel with the A–A and the allowable shear stress of these planes is 2.3 MPa, is this stress system permissible?

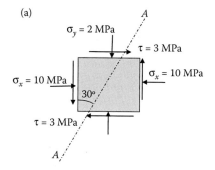

Solution

This is a question asking for calculation of principal stresses and stresses on an inclined plane. The state of stress at the point is given. The principal stresses and the shear stress on the stratified plane can be obtained by application of Equations 7.1 to 7.3 or by use of Mohr's circle method.

1. Analytical solutions: Following the sign conventions
   $\sigma_x = -10$ MPa, $\sigma_y = -2$ MPa, $\tau_{xy} = -3$ MPa (sign convention in Section 7.1.1)
   From Equations 7.2 and 7.3

$$\sigma_1 = \frac{1}{2}\left[(\sigma_x + \sigma_y) + \sqrt{(\sigma_x - \sigma_y)^2 + 4\tau_{xy}^2}\right]$$

$$= \frac{1}{2}\left[(-10 - 2) + \sqrt{(-10 + 2)^2 + 4 \times 3^2}\right] = -1.0\,\text{MPa}$$

$$\sigma_2 = \frac{1}{2}\left[(\sigma_x + \sigma_y) - \sqrt{(\sigma_x - \sigma_y)^2 + 4\tau_{xy}^2}\right]$$

$$= \frac{1}{2}\left[(-10 - 2) - \sqrt{(-10 + 2)^2 + 4 \times 3^2}\right] = -11.0\,\text{MPa}$$

$$\tan 2\theta = \frac{-2\tau}{\sigma_x - \sigma_y} = \frac{-2 \times (-3)}{-10 - (-2)} = -\frac{3}{4}$$

$\theta = -18.43°$ (clockwise from the x-axis)

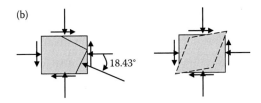

(b)

18.43°

   The above calculations suggest that one of the principal stresses acts in the direction $\theta = -18.43°$ away from the x-axis. Judging by the distortion of the element due to the applied shear stresses figure (b) above and by common sense, it is obvious that the principal stress in the direction of $\theta = -18.43°$ is 'more compressive' than the other principal stress is. Hence, the principal stress acting along $\theta = -18.43°$ is the minimum principal stress, $\sigma_2 = -11$ MPa. The maximum principal stress, $\sigma_1 = -1.0$ MPa, is in the direction perpendicular to σ_2. The principal directions at the point are shown below.

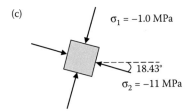

(c) $\sigma_1 = -1.0$ MPa

18.43°

$\sigma_2 = -11$ MPa

The shear stress along plane A–A is calculated from Equation 7.1.

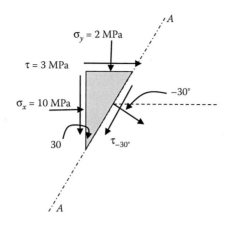

$$\tau_\theta = \left(\frac{\sigma_x - \sigma_y}{2}\right)\sin 2\theta + \tau\cos 2\theta$$

$$\tau_{-30°} = -\left(\frac{\sigma_x - \sigma_y}{2}\right)\sin 60° + \tau_{xy}\cos 60°$$

$$= -\left(\frac{-10 + 2}{2}\right)\frac{\sqrt{3}}{2} + (-3) \times \frac{1}{2} = 1.96\,\text{MPa}$$

$$\tau_{-30°} < 2.3\,\text{MPa}$$

Hence, the stress system is permissible.

2. Mohr's circle solution: According to the sign convention we have followed, the stress system shown in figure (a) above has $(\sigma_x, \tau_{xy}) = (-10\,\text{MPa}, -3\,\text{MPa})$ on the right-hand-side vertical plane and $(\sigma_x, -\tau_{xy}) = (-2\,\text{MPa}, 3\,\text{MPa})$ on the upper horizontal plane, respectively. Thus, Mohr's circle can be plotted by following the steps described in Section 7.1.3:

   a. Locate the two points with coordinates $A = (-10, -3)$ and $B = (-2, 3)$, respectively, in the $\sigma - \tau$ plane. Point A represents the right-hand side of the element and point B represents the top side of the element.

   b. Connect the two points by a straight line to establish the diameter of the circle. The intersection with the σ-axis is the average of the two normal stresses and marked 'O'.

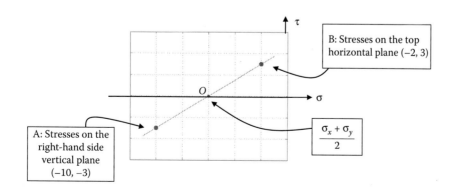

c. With its centre at 'O', draw a circle passing through the two points.

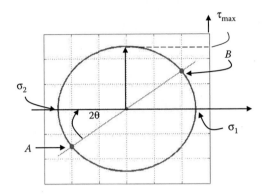

d. Measure σ coordinates of the two intersections of the circle with the σ-axis, where shear stresses are zero, to obtain the two principal stresses σ_1 and σ_2. Each division in the figure is equivalent to 2 MPa. Therefore,

$\sigma_1 = 0.5$ division $= -1$ MPa

$\sigma_2 = 5.5$ division $= -11$ MPa

e. The angle between the right-hand-side and the upper-side planes of the element (planes A and B) is 90°, while the associated points in the circle is 180° away.

This suggests that for two planes having an angle of θ, the points relative to these planes on Mohr's circle are 2θ away from each other.

The clockwise angle, 2θ, between plane A and the second principal plane is

$$\tan 2\theta = \frac{1.5 \text{ division}}{2 \text{ division}} = \frac{3}{4} = 0.75$$

$$\theta = \frac{1}{2}\text{arc tan}(0.75) = 18.43°$$

The angle between the normal stress on plane A (the vertical plane on the right-hand side of the element) and the second principal stress is, therefore, 18.43° clockwise.

f. Measure the radius of the circle to obtain the maximum shear stress

$\tau_{max} = 2.5$ division $= 5$ MPa

or take

$$\tau_{max} = \frac{(\sigma_1 - \sigma_2)}{2} = \frac{-1 + 11}{2} = 5 \text{ MPa}$$

The points associated with the maximum or minimum shear stress are 90° away from the points associated with the two principal stresses on the circle. The maximum shear stress is therefore 45° away from a principal plane.

g. Shear stress on the plane 30° clockwise away from the right-hand-side vertical planes.

Point A represents the vertical plane. The shear stress on the plane 30° away from it can be measured from the above figure, which is equal to 1.96 MPa.

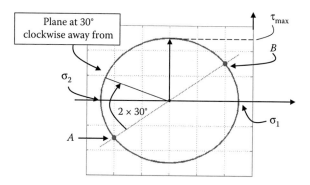

EXAMPLE 7.2

Consider the bar shown in figure below under torsional loading. Determine the principal stresses and their directions.

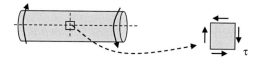

Solution

The state of stress of the shaft is determined by the shear stress caused by the torque. On the element taken, there are no normal stresses.

Thus,

$$\sigma_x = \sigma_y = 0, \quad \tau_{xy} = \tau$$

Calculating the magnitudes of σ_1 and σ_2 (Equation 7.2)

$$\sigma_1 = \frac{\sigma_1 + \sigma_2}{2} + \sqrt{\left(\frac{\sigma_1 - \sigma_2}{2}\right)^2 + \tau_{xy}^2} = \tau$$

$$\sigma_2 = \frac{\sigma_x + \sigma_y}{2} - \sqrt{\left(\frac{\sigma_x - \sigma_y}{2}\right)^2 + \tau_{xy}^2} = -\tau$$

Calculate the orientation of these principal stresses (Equation 7.3)

$$\tan 2\theta = \frac{-2\tau_{xy}}{\sigma_x - \sigma_y} = \frac{-\tau}{0} = -\infty$$

$$2\theta = -90°$$

$$\theta_1 = -45°$$

$$\theta_2 = -45° + 90° = 45°$$

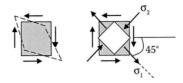

The direction of σ_1 is 45° away from the x-axis clockwise, which can be judged by the above-sketched deformation, where a tensile strain is observed in this direction.

EXAMPLE 7.3

A thin-walled cylinder has an internal diameter of 60 mm and a wall thickness of 1.5 mm. Determine the principal stresses at a point on the external surface of the generator when the cylinder is subjected to an internal pressure of 6 MPa and a torque, about its longitudinal axis, of 1.0 kN m. If the cylinder is made from plates that are welded along the 45° seams, calculate the normal and shear stresses along the seams.

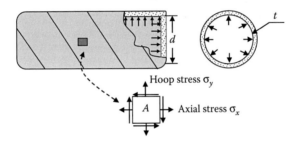

Solution

The state of stress is defined by the axial and hoop stresses caused by the internal pressure and the shear stress due to the applied torque. The pressure p acting on the end plate area A is equivalent to an axial force of

$$P_x = p \times A\text{(area)} = p \times \left(\frac{\pi d^2}{4} \right)$$

and the hoop force, P_y, is equal to

$$P_y = \frac{pd}{2}$$

For the thin-walled cylinder, the cross-sectional area can be calculated approximately by πdt.

For the infinitesimal element taken at A (figure above), the hoop stress is

$$\sigma_y = \frac{P_y}{t} = \frac{pd}{2t} = \frac{6\,\text{MPa} \times 60 \times 10^{-3}\,\text{m}}{2 \times 1.5 \times 10^{-3}\,\text{m}}$$
$$= 120\,\text{MPa}$$

The axial stress is

$$\sigma_x = \frac{P_y}{\pi dt} = \frac{pd}{4t} = \frac{6\,\text{Mpa} \times 60 \times 10^{-3}\,\text{m}}{4 \times 1.5 \times 10^{-3}\,\text{m}}$$
$$= 60\,\text{MPa}$$

Since the wall of the cylinder is thin, that is, $t \ll d$, the torsional constant is calculated approximately as

$$J = \frac{\pi}{32}(D^4 - d^4) = \frac{\pi}{32}d^4\left(\frac{D^4}{d^4} - 1\right) = \frac{\pi}{32}d^4\left[\left(1 + \frac{2t}{d}\right)^4 - 1\right] \approx \frac{\pi t}{4}d^3$$

where D is the outside diameter of the cylinder. Considering $D \approx d$, the shear stress due to the torque is then

$$\tau_{xy} = \frac{T(d/2)}{J} = \frac{1000\,\text{N m} \times 60 \times 10^{-3} \times \text{m}/2}{\pi(1.5 \times 10^{-3}\,\text{m})(60 \times 10^{-3}\,\text{m})^3/4}$$
$$= 118\,\text{MPa}$$

From Equation 7.2

$$\sigma_1 = \frac{1}{2}\left[(\sigma_x + \sigma_y) + \sqrt{(\sigma_x + \sigma_y)^2 + 4\tau_{xy}^2}\right]$$
$$= \frac{1}{2}\left[(60 + 1200) + \sqrt{(60 + 120)^2 + 4 \times 118^2}\right] = 212\,\text{MPa}$$
$$\sigma_2 = \frac{1}{2}\left[(\sigma_x + \sigma_y) + \sqrt{(\sigma_x - \sigma_y)^2 + 4\tau_{xy}^2}\right]$$
$$= \frac{1}{2}\left[(60 + 1200) + \sqrt{(60 - 120)^2 + 4 \times 118^2}\right] = -32\,\text{MPa}$$

The maximum shear stress is (Equation 7.4)

$$\tau_{max} = \frac{\sigma_x - \sigma_y}{2} = \frac{212 - (-32)}{2} = 122\,\text{MPa}$$

Along the 45° seams (Equation 7.1)

$$\sigma_\theta = \sigma_x \cos^2\theta + \sigma_y \sin^2\theta - \tau_{xy}\sin 2\theta$$
$$= 60\,\text{MPa} \times \cos^2 45° + 120\,\text{MPa} \times \sin^2 45° - 118\,\text{MPa} \times \sin 90°$$
$$= -28\,\text{MPa}$$
$$\tau_\theta = \left(\frac{\sigma_x - \sigma_y}{2}\right)\sin 2\theta + \tau_{xy}\cos 2\theta$$
$$= \left(\frac{60\,\text{MPa} - 120\,\text{MPa}}{2}\right)\sin 90° + 118\,\text{MPa} \times \cos 90°$$
$$= -30\,\text{MPa}$$

EXAMPLE 7.4

The cross section of a beam, as shown in figure below, is subjected to a bending moment of $M = 10\,kN\,m$ and a shear force of $V = 120\,kN$. Calculate the principal stresses at points 1, 2, 3 and 4, respectively.

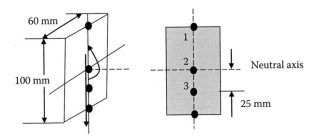

Solution

The states of stress at the points are determined by the normal stress due to the applied bending moment and the shear stress due to the applied shear force. At 1 and 4, the shear stress is zero, while at 2 the normal stress is zero. At 3 there is a combined action of normal and shear stresses. The states of stress of the four points are, respectively, as follows:

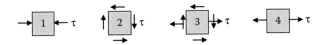

Calculating the principal stresses of the 4 points is, therefore, a direct application of Equation 7.2 to the above four states of stress.

The second moment of area of the cross section is

$$I = \frac{1}{12}bh^3 = \frac{1}{12} \times 60 \times 10^{-3}\,m \times (100 \times 10^{-3}\,m)^3 = 500 \times 10^{-8}\,m^4$$

At location 1, the normal stress due to bending (Equation 5.1)

$$\sigma = \frac{My}{I} = \frac{10 \times 10^3\,N\,m \times 50 \times 10^{-3}\,m}{500 \times 10^{-8}\,m^4} = 100 \times 10^6\,N/m^2 \text{ (compression)}$$

At 1, since there is no shear stress, the above normal stress is one of the principal stresses. Thus,

$$\sigma_1 = 0$$

$$\sigma_2 = -100\,MPa$$

At location 2, there is no normal stress (neutral axis). The shear stress is calculated by (Example 5.5)

$$\tau = \frac{3V}{2A} = \frac{3 \times 120 \times 10^3\,N}{2 \times 60 \times 100 \times 10^{-6}\,m^2} = 30 \times 10^6\,N/m^2 = 30\,MPa$$

From Equation 7.2

$$\sigma_1 = \frac{\sigma_x + \sigma_y}{2} + \sqrt{\left(\frac{\sigma_x - \sigma_y}{2}\right)^2 + \tau_{xy}^2} = 0 + \sqrt{0 + \tau^2} = \tau = 30\,\text{MPa}$$

$$\sigma_2 = \frac{\sigma_x + \sigma_y}{2} - \sqrt{\left(\frac{\sigma_x - \sigma_y}{2}\right)^2 + \tau_{xy}^2} = 0 - \sqrt{0 + \tau^2} = -\tau = -30\,\text{MPa}$$

At location 3 both normal and shear stresses exist. The normal stress due to bending is

$$\sigma = \frac{My}{I} = \frac{10 \times 10^3\,\text{N m} \times 25 \times 10^{-3}\,\text{m}}{500 \times 10^{-8}\,\text{m}^4} = 50 \times 10^6\,\text{N/m}^2 \text{ (tension)}$$

and the shear stress due to the applied shear force is (Equation 5.5)

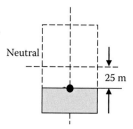

$$\tau = \frac{VS^*}{bI} = \frac{120 \times 10^3\,\text{N} \times 60 \times 25 \times 37.5 \times 10^{-9}\,\text{m}^3}{60 \times 10^{-3}\,\text{m} \times 500 \times 10^{-8}\,\text{m}^4}$$

$$= 22.5 \times 10^6\,\text{N/m}^2$$

From Equation 7.2

$$\sigma_1 = \frac{\sigma_x + \sigma_y}{2} + \sqrt{\left(\frac{\sigma_x - \sigma_y}{2}\right)^2 + \tau_{xy}^2}$$

$$= \frac{50 \times 10^6\,\text{Pa} + 0}{2} + \sqrt{\left(\frac{50 \times 10^6\,\text{Pa} - 0}{2}\right)^2 + \tau^2} = 58.6\,\text{MPa}$$

$$\sigma_2 = \frac{\sigma_x + \sigma_y}{2} - \sqrt{\left(\frac{\sigma_x - \sigma_y}{2}\right)^2 + \tau_{xy}^2}$$

$$= \frac{50 \times 10^6\,\text{Pa} + 0}{2} - \sqrt{\left(\frac{50 \times 10^6\,\text{Pa} - 0}{2}\right)^2 + \tau^2} = 8.6\,\text{MPa}$$

At location 4, there is no shear stress and the normal stress due to bending at the point is one of the principal stresses

$$\sigma = \frac{My}{I} = \frac{10 \times 10^3\,\text{N m} \times 50 \times 10^{-3}\,\text{m}}{500 \times 10^{-8}\,\text{m}^4} = 100 \times 10^6\,\text{N/m}^2 \text{ (tension)}$$

Thus,

$\sigma_1 = 100$ MPa

$\sigma_2 = 0$

7.4 CONCEPTUAL QUESTIONS

1. What is meant by 'state of stress' and why is it important in stress analysis?
2. The state of stress at a point of a material is completely determined when
   a. The stresses on three mutually perpendicular planes are specified
   b. The stresses on two mutually perpendicular planes are specified
   c. The stresses on an arbitrary plane are specified
   d. None of the above statement is correct
3. Can both the square elements shown in figure below be used to represent the state of stress at the point of the beam? If yes, which one do you prefer to use and why?

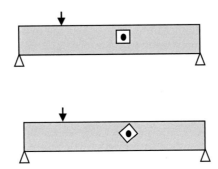

4. Consider a shaft with a constant circular section. If it is subjected to torsion only, at any point within the shaft, normal stress is always zero. (Y/N)
5. What is meant by 'principal stresses', and what is its importance?
6. How many principal stresses are there at a point in a two-dimensional stress system?
7. What is meant by 'principal planes', and what is the value of shear stress on these planes?
8. If the principal stresses at a point are known, how can the maximum shear stress at the point be calculated?
9. On the maximum shear stress plane, what values do the normal stresses take?
10. What is the angle between a principal plane and a maximum shear stress plane? Use Mohr's circle to illustrate this.
11. From a Mohr's circle, why do the intersections with the horizontal axis represent the principal stresses at a point?
12. From a Mohr's circle, why does the radius of the circle represent the maximum shear stress at a point?
13. Use Mohr's circle to illustrate the equality of shear stresses on two planes that are perpendicular to each other.
14. The state of stress at a point is shown in figure below. The angle between the x-axis and the maximum normal stress is likely to be in the direction of

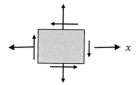

a. 13.5°
b. −96.5°
c. 76.5°
d. −13.5°

(Positive angle is defined as anticlockwise from the x-axis.)

15. In Equation 7.2, what conclusion can you draw from the sum of σ_1 and σ_2?

7.5 MINI TEST

PROBLEM 7.1

Among the three states of stress shown in figure below, which of the two are equivalent?

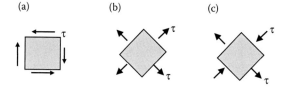

PROBLEM 7.2

A thin-walled hollow sphere of radius R with uniform thickness t is subjected to an internal pressure p. Determine and discuss the state of stress of a point on the outside surface. Is there any shear stress acting at the point? Explain why.

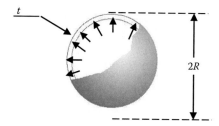

PROBLEM 7.3

Match the states of stress shown in Figure (b) below with the points shown on the two beams in Figure (a) below.

(a)

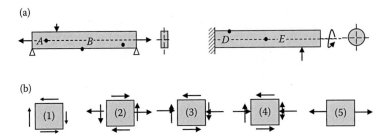

(b)

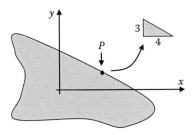

PROBLEM 7.4

At point P on the free edge of a plate, oriented as shown in figure below, the maximum shear stress at the point is 4000 kN/m^2. Use Mohr's circle to find the principal stresses and σ_x at the point in the x–y system shown in the figure:

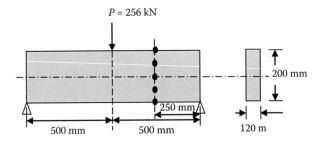

PROBLEM 7.5

A simply supported beam of rectangular cross section is loaded as shown in figure below. Determine the states of stress of points 1–5 that are equally spaced across the depth of the beam. Calculate also the principal stresses at these points.

$P = 256$ kN

200 mm

250 mm

120 m

500 mm   500 mm

Chapter 8

Complex strains and strain gauges

From Chapter 7, at a point within a three-dimensional material, there are usually six independent stresses as shown in Figure 8.1, that is,

- Three normal stresses: σ_x, σ_y and σ_z
- Three shear stresses: τ_{xy}, τ_{xz} and τ_{yz}

At the same point, due to the action of these stresses, there exist six independent strains, that is,

- Three normal strains: ε_x, ε_y and ε_z
- Three shear strains: γ_{xy}, γ_{yz} and γ_{yz}

For linearly elastic and isotropic materials, the six stresses and the six strains satisfy Hooke's law as follows:

$$\varepsilon_x = \frac{\sigma_x}{E} - v\frac{\sigma_y}{E} - v\frac{\sigma_z}{E}$$
$$= \frac{1}{E}[\sigma_x - v(\sigma_y + \sigma_z)]$$

$$\varepsilon_y = \frac{\sigma_y}{E} - v\frac{\sigma_x}{E} - v\frac{\sigma_z}{E}$$
$$= \frac{1}{E}[\sigma_y - v(\sigma_x + \sigma_z)]$$

$$\varepsilon_z = \frac{\sigma_z}{E} - v\frac{\sigma_x}{E} - v\frac{\sigma_y}{E}$$
$$= \frac{1}{E}[\sigma_z - v(\sigma_x + \sigma_y)]$$

(8.1)

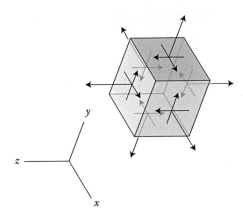

Figure 8.1 Three-dimensional stresses.

$$\gamma_{xy} = \frac{\tau_{xy}}{G}$$

$$\gamma_{xz} = \frac{\tau_{xz}}{G} \qquad (8.2)$$

$$\gamma_{yz} = \frac{\tau_{yz}}{G}$$

where E, G and v are, respectively, Young's modulus, shear modulus and Poisson's ratio of the material.

In a two-dimensional stress system, where one of the normal stresses, for example, the normal stress in the z direction, is zero, or in most cases negligible, and the shear stresses relative to this direction are also zero, the generalised Hooke's law of Equations 8.1 and 8.2 is reduced to the following:

- Strains in terms of stresses:

$$\varepsilon_x = \frac{\sigma_x}{E} - v\frac{\sigma_y}{E} = \frac{1}{E}[\sigma_x - v\sigma_y]$$

$$\varepsilon_y = \frac{\sigma_y}{E} - v\frac{\sigma_x}{E} = \frac{1}{E}[\sigma_y - v\sigma_x] \qquad (8.3)$$

$$\gamma_{xy} = \frac{\tau_{xy}}{G}$$

- Stresses in terms of strains:

$$\sigma_x = \frac{E}{1-v^2}[\varepsilon_x + v\varepsilon_y]$$

$$\sigma_y = \frac{E}{1-v^2}[\varepsilon_y + v\varepsilon_x] \qquad (8.4)$$

$$\tau_{xy} = G\gamma_{xy}$$

Equations 8.3 and 8.4 represent a two-dimensional complex strain system (Figure 8.2) that is directly related to the two-dimensional complex stress system discussed in Section 7.1.

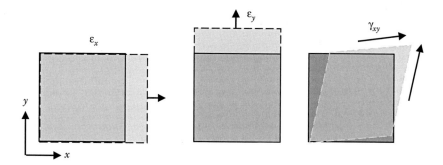

Figure 8.2 Two-dimensional strains.

The collective information of ε_x, ε_y and γ_{xy} at a point within a two-dimensional system is termed as *state of strain* at that point.

In Figure 8.2, if the shear strain, γ_{xy}, is zero (no distortion), the two normal strains are called *principal strains* and are designated as ε_1 and ε_2, where $\varepsilon_1 > \varepsilon_2$. The two principal strains are associated with the two principal stresses (Figure 8.3) discussed in Chapter 7.

For the principal stresses and principal strains shown in Figure 8.3, Hooke's law (8.3 and 8.4) is reduced to the following:

- Strains in terms of stresses:

$$\varepsilon_1 = \frac{\sigma_1}{E} - v\frac{\sigma_2}{E} = \frac{1}{E}[\sigma_1 - v\sigma_2]$$
$$\varepsilon_2 = \frac{\sigma_2}{E} - v\frac{\sigma_1}{E} = \frac{1}{E}[\sigma_2 - v\sigma_1]$$
(8.5)

- Stresses in terms of strains:

$$\sigma_1 = \frac{E}{1 - v^2}[\varepsilon_1 + v\varepsilon_2]$$
$$\sigma_2 = \frac{E}{1 - v^2}[\varepsilon_2 + v\varepsilon_1]$$
(8.6)

where both shear stress and shear strain are zero.

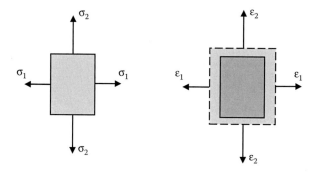

Figure 8.3 Two-dimensional principal stresses and strains.

- When a structure is subjected to external loads, the state of strains is, in general, different at different points within the structure.
- At a point in a structure, the strain in one direction is usually different to the strain in a different direction.
- State of strains (Figure 8.2) shows two normal strains and a shear strain at a point. When a normal strain is known, it means that not only the magnitude, but also the direction of the strain is known. The shear strain is always in consistence with the shear stresses at the point.
- The maximum and minimum normal strains are called *principal strains*. They are, respectively, associated with the maximum and minimum principal stresses.
- The shear strain relative to the maximum shear stress is the maximum shear strain.

8.1 STRAIN ANALYSIS

In most cases, the purpose of a strain analysis is to carry out a stress analysis. This is particularly true when the analysis is based on experiments. Usually, it is much easier and more straightforward to measure strains (deformation) than stresses by experiments. The measured strains can then be converted into stresses on the basis of Hooke's law. Like the stress analysis discussed in Chapter 7, the strain analysis here again deals with the following two issues:

- To determine the strain in relation to an important direction based on the measured strains aligning with a chosen coordinate system.
- To determine the maximum normal strain or maximum shear strain at a point. These strains may not necessarily align with the chosen coordinate directions.

1. Strains transformation: The two states of strains are taken at the same point of a material (Figure 8.4). By comparing Figure 8.4 with Figure 7.5, the strains in the x–y and the x'–y' coordinate systems should have a similar relationship to that for stresses. Considering the fact that the shear strain, γ_{xy}, is related to a pair of shear stresses, τ_{xy}

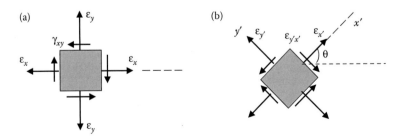

Figure 8.4 Strain transformation.

and τ_{yx}, only $\gamma_{xy}/2$ will appear in the equation as an equivalent to the shear stress τ_{xy}. Thus, the strains in an arbitrary direction can be obtained by replacing σ with ε and t with $\gamma/2$, respectively in Equation 7.1.

$$\varepsilon_\theta = \varepsilon_{x^1} = \varepsilon_x \cos^2\theta + \varepsilon_y \sin^2\theta - \frac{\gamma_{xy}}{2}\sin 2\theta$$

$$= \frac{\varepsilon_x + \varepsilon_y}{2} + \frac{\varepsilon_x - \varepsilon_y}{2}\cos 2\theta - \frac{\gamma_{xy}}{2}\sin 2\theta \qquad (8.7)$$

$$\frac{\gamma_\theta}{2} = \frac{\gamma_{x'y'}}{2} = \left(\frac{\varepsilon_x - \varepsilon_y}{2}\right)\sin 2\theta + \frac{\gamma_{xy}}{2}\cos 2\theta$$

or

$$\gamma_\theta = \gamma_{x'y'} = (\varepsilon_x - \varepsilon_y)\sin 2\theta + \gamma_{xy}\cos 2\theta$$

where an anticlockwise angle from the x-axis is defined as positive.

2. Principal strains and their directions: Following the same argument for the above strain transformation, the principal strains and their directions can be deducted from the principal stress equations (Equations 7.2 and 7.3).

$$\varepsilon_1 = \frac{1}{2}\left[(\varepsilon_x + \varepsilon_y) + \sqrt{(\varepsilon_x - \varepsilon_y)^2 + \gamma_{xy}^2}\right]$$

$$\varepsilon_1 = \frac{1}{2}\left[(\varepsilon_x + \varepsilon_y) - \sqrt{(\varepsilon_x - \varepsilon_y)^2 + \gamma_{xy}^2}\right]$$

$$\tan 2\theta = \frac{-\gamma_{xy}}{\varepsilon_x - \varepsilon_y} \qquad (8.8)$$

$$\theta = \frac{1}{2}\tan^{-1}\left(\frac{-\gamma_{xy}}{\varepsilon_x - \varepsilon_y}\right)$$

3. Maximum shear strain: Again with the same argument, from Equations 7.4 and 7.5.

$$\frac{\gamma_{max}}{2} = \frac{\varepsilon_x - \varepsilon_y}{2}$$

or

$$\gamma_{max} = \varepsilon_1 - \varepsilon_2$$

$$\frac{\gamma_{max}}{2} = \frac{1}{2}\sqrt{(\varepsilon_x - \varepsilon_y)^2 + \gamma_{xy}^2} \qquad (8.9)$$

or

$$\gamma_{max} = \sqrt{(\varepsilon_x - \varepsilon_y)^2 + \gamma_{xy}^2}$$

8.2 STRAIN MEASUREMENT BY STRAIN GAUGES

Experimental strain and stress analysis is an important tool in structural testing and design, where deformations or strains are usually measured and then converted into stresses. Strain gauges are by far the most commonly adopted method of measuring strains.

When a strain gauge is mounted on the surface of a structural member subject to deformation, the gauge deforms (elongated or contracted) with the structure. The electrical resistance of the gauge changes as the gauge deforms. This change is recorded by the strain meter and subsequently converted into the real normal strain in the longitudinal direction of the gauge at the point of measurement (Figure 8.5).

- If the normal strain in a particular direction at a surface point of a material is required, a single strain gauge is mounted along the required direction at the point.
- If the directions of principal stresses at a surface point are known in advance, strain measurements in the directions of the principal stresses are needed to obtain the principal strains and stresses.
- If the directions of principal stresses at a surface point are unknown in advance, a strain gauge rosette, which is an arrangement of three closely positioned gauges and separately oriented grids, is needed to measure the normal strains along the three different directions that are required to determine the principal strains and stresses.

The strain gauge rosette shown in Figure 8.6 is the most commonly used type of rosette and is called *rectangular rosette*.

When a rectangular rosette is mounted on the surface point of a material, the strains along the a, b and c directions of the material at the point are measured as ε_a, ε_b and ε_c, respectively. These measurements are then introduced into the following equations to determine the principal strains:

$$\varepsilon_1 = \frac{1}{2}(\varepsilon_a + \varepsilon_c) + \frac{1}{\sqrt{2}}\sqrt{(\varepsilon_a - \varepsilon_b)^2 + (\varepsilon_c - \varepsilon_b)^2}$$

$$\varepsilon_2 = \frac{1}{2}(\varepsilon_a + \varepsilon_c) - \frac{1}{\sqrt{2}}\sqrt{(\varepsilon_a - \varepsilon_b)^2 + (\varepsilon_c - \varepsilon_b)^2} \tag{8.10}$$

The directions of the principal strains can be determined by

$$\tan 2\theta = \frac{2\varepsilon_b - \varepsilon_a - \varepsilon_c}{\varepsilon_a - \varepsilon_c} \tag{8.11}$$

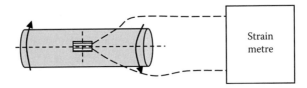

Figure 8.5 Torsion of a circular bar.

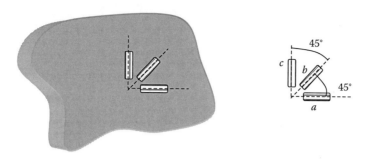

Figure 8.6 Strain measurement using strain gauge rosette.

The principal stresses at the point are found by introducing the principal strains of Equation 8.9 into Equation 8.6.

8.3 KEY POINTS REVIEW

8.3.1 Complex strain system

- At a point in a material, there are six independent strain components, including three normal strains and three shear strains.
- In a two-dimensional case, there are three independent strains, two normal strains and one shear strain at a point of a material.
- A strain usually varies from point to point.
- Strains are dimensionless.
- Principal strains are normal strains and include both maximum and minimum strains at a point.
- In a three-dimensional system, there are three principal strains; while in a two-dimensional system, there are two principal strains.
- Principal strains and principal stresses have the same directions, and are related by Hooke's law.

8.3.2 Strain measurement by strain gauges

- Strain gauges can only measure local strains at a point.
- Strain gauges measure the strains in the plane of the gauge.
- A strain gauge measures the normal strain in the longitudinal direction of the gauge.
- If the directions of principal stresses are known, only two strain gauges are required to determine the principal strains at a point.
- A strain gauge rosette consisting of at least three gauges is sufficient to measure the principal strains at any surface point of a material.

8.4 EXAMPLES

EXAMPLE 8.1

A thin-walled cylinder is subjected to an internal pressure as shown in figure below. At a point on the outside surface of the generator, the two strain gauges recorded are, respectively, $\varepsilon_a = 254 \times 10^{-6}$ and $\varepsilon_b = 68 \times 10^{-6}$. Determine the principal stresses at the surface point. Young's modulus and Poisson's ratio of the cylinder are, respectively, $E = 210$ GPa and $v = 0.28$.

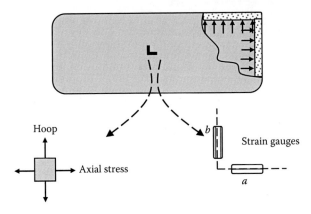

Solution

Due to the symmetric deformation and absence of shear stress on the cross section, the state of stress can be defined by the axial and hoop stresses caused by the internal pressure. The two stresses are the principal stresses. The two strain measurements from the strain gauge are, therefore, the two principal strains. Thus, a straightforward application of Equation 8.6 yields the principal stresses at the point.

Let

$$\varepsilon_1 = \varepsilon_a \quad \text{and} \quad \varepsilon_2 = \varepsilon_b$$

$$\sigma_1 = \frac{E}{1-v^2}[\varepsilon_1 + v\varepsilon_2] = \frac{210\,\text{GPa}}{1-0.28^2}[254 \times 10^{-6} + 0.28 \times 68 \times 10^{-6}]$$
$$= 0.062\,\text{GPa} = 62\,\text{MPa}$$
$$\sigma_2 = \frac{E}{1-v^2}[\varepsilon_2 + v\varepsilon_1] = \frac{210\,\text{GPa}}{1-0.28^2}[68 \times 10^{-6} + 0.28 \times 254 \times 10^{-6}]$$
$$= 0.032\,\text{GPa} = 32\,\text{MPa}$$

The two principal stresses are, respectively, 62 and 32 MPa.

EXAMPLE 8.2

At a surface point of a beam, a rectangular strain gauge rosette (Figure 8.6) recorded the following strains:

$$\varepsilon_a = -270 \times 10^{-6}, \quad \varepsilon_b = -550 \times 10^{-6}, \quad \varepsilon_c = 80 \times 10^{-6}$$

If the gauges 'a' and 'c' are in line with and perpendicular to the axis of the beam, calculate the principal stresses at the point and their direction. Take $E = 200$ GPa and $v = 0.3$.

Solution

This question requires direct application of Equations 8.9 and 8.10 to compute principal strains and their directions, and Equation 8.6 to compute principal stresses. For isotropic materials, the directions of principal stresses coincide with the directions of principal strains.

From Equation 8.9

$$\varepsilon_1 = \frac{1}{2}(\varepsilon_a + \varepsilon_c) + \frac{1}{\sqrt{2}}\sqrt{(\varepsilon_a - \varepsilon_b)^2 + (\varepsilon_c - \varepsilon_b)^2}$$

$$= \frac{1}{2}(-270 + 80) \times 10^{-6} + \frac{10^{-6}}{\sqrt{2}}\sqrt{(-270 + 550)^2 + (80 + 550)^2}$$

$$= 392.49 \times 10^{-6}$$

$$\varepsilon_2 = \frac{1}{2}(\varepsilon_a + \varepsilon_c) - \frac{1}{\sqrt{2}}\sqrt{(\varepsilon_a - \varepsilon_b)^2 + (\varepsilon_c - \varepsilon_b)^2}$$

$$= \frac{1}{2}(-270 + 80) \times 10^{-6} - \frac{10^{-6}}{\sqrt{2}}\sqrt{(-270 + 550)^2 + (80 + 550)^2}$$

$$= -582.49 \times 10^{-6}$$

From Equation 8.6

$$\sigma_1 = \frac{E}{1 - v^2}[\varepsilon_1 + v\varepsilon_2] = \frac{200 \times 10^9 \, \text{Pa}}{1 - 0.3^2}[392.49 \times 10^{-6} + 0.3 \times (-582.49 \times 10^{-6})]$$

$$= 47.86 \, \text{MPa}$$

$$\sigma_2 = \frac{E}{1 - v^2}[\varepsilon_2 + v\varepsilon_1] = \frac{200 \times 10^9 \, \text{Pa}}{1 - 0.3^2}[-582.49 \times 10^{-6} + 0.3 \times 392.49 \times 10^{-6}]$$

$$= -102.14 \, \text{MPa}$$

From Equation 8.10

$$\tan 2\theta = \frac{2\varepsilon_b - \varepsilon_a - \varepsilon_c}{\varepsilon_a - \varepsilon_c} = \frac{2 \times (-550) - (-270) - 80}{-270 - 80}$$

$$= 2.6$$

Thus,

$$2\theta = 68.8°$$

$$\theta = 34.4°$$

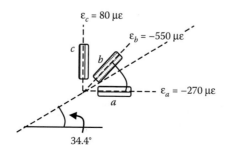

In order to determine which principal stress (strain) is associated with the above angle, the conceptual analysis below can be followed. We know that one of the principal stresses (strains) is in the direction of 34.4° from gauge 'a', and the normal strain in this direction must be compressive (negative), judging by the compressive strains of gauges 'a' and 'b'. The principal stress (strain) in this direction must be compressive and is the minimum principal stress (strain). Thus,

$$\sigma_1 = 47.86\,\text{MPa}; \quad \theta_1 = \theta + 90° = 34.4° + 90° = 124.4°$$
$$\sigma_2 = -102.14\,\text{MPa}; \quad \theta_2 = \theta = 34.4°$$

EXAMPLE 8.3

At a point on the surface of the generator (figure below), the readings of the rectangular strain gauge rosette are $\varepsilon_a = 100 \times 10^{-6}$, $\varepsilon_b = -200 \times 10^{-6}$ and $\varepsilon_c = -300 \times 10^{-6}$. The gauges 'a' and 'c' are in line with and perpendicular to the axis of the beam. Determine the principal stresses at the surface point and the magnitudes of the applied twist moment and the axial force. Young's modulus and Poisson's ratio of the cylinder are, respectively, $E = 70$ GPa and $v = 0.3$.

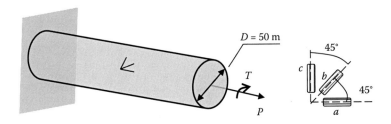

Solution

The principal stresses can be easily computed by Equations 8.9 and 8.6. To determine the applied axial force, the relationship between the principal stresses and the normal stresses on the cross section at the point must be established, from which the normal stress and then the axial force can be calculated. To determine the applied torque, the relationship between the principal stresses and the shear stresses on the cross section is required since the shear stress is directly related to the applied torque.

1. Principal strains (Equation 8.9):

$$\varepsilon_1 = \frac{1}{2}(\varepsilon_a + \varepsilon_c) + \frac{1}{\sqrt{2}}\sqrt{(\varepsilon_a - \varepsilon_b)^2 + (\varepsilon_c - \varepsilon_b)^2}$$

$$= \frac{1}{2}(1000 - 300) \times 10^{-6} + \frac{10^{-6}}{\sqrt{2}}\sqrt{(1000 + 200)^2 + (-300 + 200)^2}$$

$$= 1202 \times 10^{-6}$$

$$\varepsilon_2 = \frac{1}{2}(\varepsilon_a + \varepsilon_c) - \frac{1}{\sqrt{2}}\sqrt{(\varepsilon_a - \varepsilon_b)^2 + (\varepsilon_b - \varepsilon_c)^2}$$

$$= \frac{1}{2}(1000 - 300) \times 10^{-6} - \frac{10^{-6}}{\sqrt{2}}\sqrt{(1000 + 200)^2 + (-300 + 200)^2}$$

$$= -502 \times 10^{-6}$$

2. Principal stresses (Equation 8.6):

$$\sigma_1 = \frac{E}{1-v^2}[\varepsilon_1 + v\varepsilon_2] = \frac{70,000}{1-0.3^2}(1,202 - 0.3 \times 502) \times 10^{-6}$$

$$= 80.09 \, \text{N/mm}^2$$

$$\sigma_2 = \frac{E}{1-v^2}[\varepsilon_2 + v\varepsilon_1] = \frac{70,000}{1-0.3^2}(-502 + 0.3 \times 1,202) \times 10^{-6}$$

$$= -10.9 \, \text{N/mm}^2$$

3. Stresses (σ_x and τ_{xy}) on the cross section: Taking the axial direction as the x direction, the state of stress at the surface point is as follows:

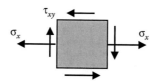

where the normal stress, σ_x, is related to the axial force F and the shear stress, τ_{xy}, is related to the twist moment T. In this case the hoop stress is zero, that is, $\sigma_y = 0$. From Equation 7.2, the sum of the two principal stresses yields

$\sigma_1 + \sigma_2 = \sigma_x + \sigma_y$ (This equation holds for any two-dimensional problem.)

Thus,

$$\sigma_x = \sigma_1 + \sigma_2 = 80.09 - 10.9 = 70 \, \text{N/mm}^2 (\sigma_y = 0)$$

On the cross section (Equation 2.1)

$$\sigma_x = \frac{F}{A}$$

$$F = \sigma_x A = 70 \times \frac{\pi}{4} \times 50^2 = 137.4 \, \text{kN}$$

Again from Equation 7.2, the difference between the two principal stresses is

$$\sigma_1 - \sigma_2 = \sqrt{(\sigma_x - \sigma_y)^2 + 4\tau_{xy}^2}$$

In this case, $\sigma_y = 0$. Thus,

$$\tau_{xy} = \frac{1}{2}\sqrt{(\sigma_1 - \sigma_2)^2 - \sigma_x^2} = \frac{1}{2}\sqrt{(80.09 - 10.9)^2 - 70^2}$$

$$= 19.7 \, \text{N/mm}^2$$

On the cross section (Equation 3.2)

$$\tau_{xy} = \frac{Tr}{J}$$

$$T = \frac{\tau_{xy}J}{r} = \frac{29.7 \, (\pi/32) \times 50^4}{25} = 0.7 \, \text{kN m}$$

The axial force and twist moment applied at the free end are, respectively, 137.4 kN and 0.7 kN m.

EXAMPLE 8.4

A cantilever beam of solid circular cross section has a diameter 100 mm and is 1 m long. An arm of length d is attached to its free end. The arm carries a load W as shown in figure below. On the beam's upper surface, halfway along its length and coinciding with its vertical plane of symmetry is a rectangular strain gauge rosette, which gave the following readings for particular values of W and d:

$$\varepsilon_a = 1500 \times 10^{-6}, \quad \varepsilon_b = -300 \times 10^{-6}, \quad \varepsilon_c = -450 \times 10^{-6}$$

If gauges 'a' and 'c' are in line with and perpendicular to the axis of the beam, calculate the values of W and d. Take $E = 200,000$ N/mm² and $v = 0.3$.

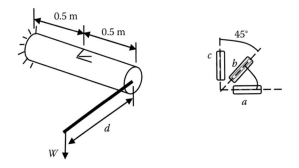

Solution

Again, a direct application of Equations 8.9 and 8.6 can provide the principal strains and principal stresses at the point. To determinate the load W and then lever arm d, the bending moment and twist moment due to W and d on the cross section at the location where the strain gauge rosette is mounted must be calculated. The relationships between the bending and twist moments and the normal and shear stresses on the section are then established, from which the values of W and d are finally determined.

1. Principal strains: Substituting values of ε_a, ε_b and ε_c in Equation 8.9 gives

$$
\begin{aligned}
\varepsilon_1 &= \frac{1}{2}(\varepsilon_a + \varepsilon_c) + \frac{1}{\sqrt{2}}\sqrt{(\varepsilon_a - \varepsilon_b)^2 + (\varepsilon_c - \varepsilon_b)^2} \\
&= \frac{1}{2}(1500 - 450) \times 10^{-6} + \frac{10^{-6}}{\sqrt{2}}\sqrt{(1500 + 300)^2 + (-450 + 300)^2} \\
&= 1802.1 \times 10^{-6}
\end{aligned}
$$

$$
\begin{aligned}
\varepsilon_2 &= \frac{1}{2}(\varepsilon_a + \varepsilon_c) - \frac{1}{\sqrt{2}}\sqrt{(\varepsilon_a - \varepsilon_b)^2 + (\varepsilon_c - \varepsilon_b)^2} \\
&= \frac{1}{2}(1500 - 450) \times 10^{-6} - \frac{10^{-6}}{\sqrt{2}}\sqrt{(1500 + 300)^2 + (-450 + 300)^2} \\
&= 752.2 \times 10^{-6}
\end{aligned}
$$

2. Principal stresses: Now from Equation 8.6

$$\sigma_1 = \frac{E}{1 - v^2}(\varepsilon_1 + v\varepsilon_2)$$

$$= \frac{200,000}{1 - 0.3^2}(1,802.2 - 0.3 \times 752.2) \times 10^{-6}$$

$$= 346.5 \, \text{N/mm}^2$$

$$\sigma_2 = \frac{E}{1 - v^2}(\varepsilon_2 + v\varepsilon_1)$$

$$= \frac{200,000}{1 - 0.3^2}(-752.2 + 0.3 \times 1,802.2) \times 10^{-6}$$

$$= -46.5 \, \text{N/mm}^2$$

3. Stresses (σ_x and τ_{xy}) on the cross section: Taking the axial direction as the x direction, the state of stress at the surface point is as follows:

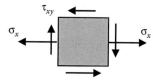

where the normal stress, σ_x, is related to the bending moment caused by W, and the shear stress, τ_{xy}, is related to the twist moment caused by W and d. The hoop stress is zero, that is, $\sigma_y = 0$.

From Equation 7.2, the sum of the two principal stresses yields

$$\sigma_1 + \sigma_2 = \sigma_x + \sigma_y$$

Thus,

$$\sigma_x = \sigma_1 + \sigma_2 = 346.5 - 46.5 = 300 \, \text{N/mm}^2$$

On the cross section, the bending moment is $M = W \times 0.5 \, \text{m}$
From Equation 5.1

$$\sigma_x = \frac{My}{I} = \frac{W \times 500 \, \text{mm} \times 50 \, \text{mm}}{\pi \times 100^4 / 64 \, \text{mm}^4} = 300 \, \text{N/mm}^2$$

$$W = 58.9 \, \text{kN}$$

Again from Equation 7.2, the difference between the two principal stresses is

$$\sigma_1 - \sigma_2 = \sqrt{(\sigma_x - \sigma_y)^2 + 4\tau_{xy}^2}$$

$$\tau_{xy} = \frac{1}{2}\sqrt{(\sigma_1 - \sigma_2)^2 - \sigma_x^2} = \frac{1}{2}\sqrt{(346.5 + 45.5)^2 - 300^2}$$

$$= 126.9 \, \text{N/mm}^2$$

On the cross section, the twist moment is

$$T = W \times d = 58.9 \times d$$

From Equation 3.2

$$\tau_{xy} = \frac{Tr}{J} = \frac{W \times d \times r}{J} = \frac{58.9 \times d \times 50}{\pi \times 100^4/32}$$
$$= 126.9\,\text{N/mm}^2$$
$$d = 423\,\text{mm}$$

The load applied at the arm and the distance is, respectively, 58.9 kN and 423 mm.

8.5 CONCEPTUAL QUESTIONS

1. What is meant by 'state of strain' and why is it important in stress analysis?
2. What is meant by 'principal strains' and what are its importance?
3. How many principal strains are there at a point in a two-dimensional stress system?
4. What strain can be measured directly by a strain gauge?
5. At a surface point of a material if the directions of two principal stresses are known, how many strain gauges are needed to measure the principal strains? Describe how the principal stresses can be calculated.
6. At a surface point of a material if the directions of two principal stresses are unknown, how many strain gauges are needed to measure the principal strains? Describe how the principal strains and principal stresses can be calculated.
7. Discuss how Young's modulus and Poisson's ratio can be measured by the strain gauges technique.
8. Can strain gauges be used to measure shear strain? Explain how this can be done.

8.6 MINI TEST

PROBLEM 8.1

A bar of circular section is subjected to torsion only (figure below). A single strain gauge is used to determine the shear stress on the cross section. What is the best orientation in which the strain gauge should be mounted on the bar?

PROBLEM 8.2

In the simple tension test shown in figure below, if the two strain gauges recorded ε_a and ε_b, respectively, describe how Young's modulus and Poisson's ratio can be determined.

PROBLEM 8.3

The single strain gauge on the top surface of the cantilever of rectangular section shown in figure below recorded a longitudinal strain ε_0. If load W is applied to the free end of the cantilever, determine the longitudinal stress at the location and the applied load W.
   Young's modulus of the material is E.

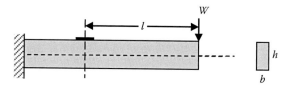

PROBLEM 8.4

The strain readings from a rectangular strain gauge rosette are given below:

$$\varepsilon_a = 1000 \times 10^{-6}, \quad \varepsilon_b = 200 \times 10^{-6}, \quad \varepsilon_c = -600 \times 10^{-6}$$

which are the strains recorded in the 0°, 45° and 90° directions, respectively. Find the principal strains and their orientations. Find also the maximum shear strain.

PROBLEM 8.5

The cantilever column of rectangular cross section shown in figure below is subjected to a wind pressure of intensity ω and a compressive force F. A rectangular strain gauge rosette located at a surface point Q is positioned on the centroidal axis 2 m downwards from the free end. The rosette recorded the following strain readings:

$$\varepsilon_a = -222 \times 10^{-6}, \quad \varepsilon_b = -213 \times 10^{-6}, \quad \varepsilon_c = +45 \times 10^{-6}$$

where gauges 'a', 'b' and 'c' are in line with, at 45° to and perpendicular to the centroidal axis of the column, respectively. Calculate the direct stress and the shear stress at point Q in the horizontal plane and hence the compressive force F and the transverse pressure ω. $E = 31,000$ N/mm^2, $v = 0.2$.

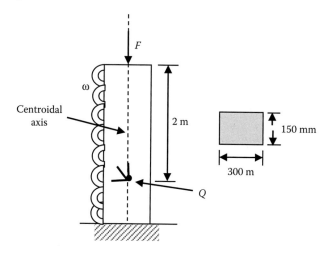

Chapter 9

Theories of elastic failure

When a structural component is subjected to increasing loads, it will eventually fail. Failure is a condition that prevents a structure from performing the intended task.

In practical applications, failure can be defined as

- Fracture with very little yielding
- Permanent deformation

The resistance of a material to failure is called *strength*. It is comparatively easy to determine the strength or the point of failure of a component subjected to a single tensile force. For example, for the bars shown in Figure 9.1a, the material fractures when the principal stress approaches the fracture stress in a tensile test. This failure mode occurs normally for bars made of brittle materials, such as cast iron, and is best demonstrated by such a bar subjected to torsion (Figure 9.1b), where the maximum principal stress acts in the direction 45° to the longitudinal axis.

However, if the material of the bars shown in Figure 9.1 is replaced by a ductile material, for example, mild steel, the failure mode is significantly different. For the bar in tension (Figure 9.2a), the bar breaks after undergoing permanent local deformation (yielding). Compared with the bar subjected to torsion in Figure 9.1, the failure surface of the ductile bar is almost normal to the longitudinal axis (Figure 9.2b), where the shear stress has reached the shear strength of the material.

From the above simple examples, it can be concluded that material property is a predominant factor that must be considered in failure analysis.

When a material is subjected to a combination of tensile, compressive and shear stresses, it is far more complicated to determine analytically whether or not the material has failed, and how the material will fail. Most of the information on yielding or fracture of materials subjected to complex stress system comes from practical design experience, experimental evidence and interpretation of them. Investigations on the information enable a formulation of theories of failure to be established for materials subjected to complex stresses.

The establishment of failure criteria for complex stress system is based on the extension of the concept of failure criteria for a material subjected to a uniaxial stress to materials subjected to combined stresses. The basis for this extension is the introduction of an equivalent stress (σ_{eq}) that represents a combined action of the stress components of a complex stress system. It is assumed that failure will occur in a material subjected to a complex stress system when this equivalent stress reaches a limiting value (σ_{Yield}) that is equal to the failure (yield or fracture) stress of the same material subjected to simple tension.

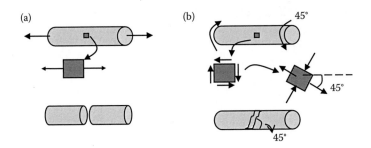

Figure 9.1 Brittle failure of a circular bar.

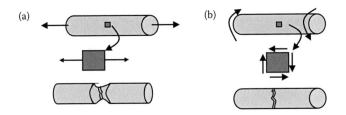

Figure 9.2 Ductile failure of a circular bar.

Materials can be broadly separated into ductile and brittle materials. Examples of ductile materials include mild steel, copper and the like. Cast iron and concrete are typical examples of brittle materials. Brittle materials experience little deformation prior to failure, and failure is generally sudden. A ductile material is considered to have failed when a marked plastic deformation has begun. A number of theories of elastic failure are widely recognised, including the following:

- Maximum principal stress theory
- Maximum shear stress theory (Tresca theory)
- Maximum distortional energy density theory (von Mises theory)

The selection of failure criteria usually depends on a number of aspects of a particular design, including material properties, state of stress, temperature and design philosophy. It may be possible that there exist several failure criteria that are applicable to a material in a particular design. However, in most cases, failure criteria are classified as applicable to brittle or ductile materials.

9.1 MAXIMUM PRINCIPAL STRESS CRITERION

This criterion assumes that principal stress is the driving factor that causes failure of materials. According to this criterion, the following comparison has been made between a complex stress system and a simple tension/compression test.

Two-dimensional complex state of stress

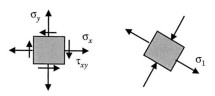

Simple tension/compression *at failure*

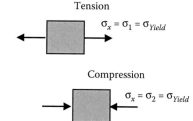

From Equation 7.2, the principal stresses are

$$\sigma_1 = \frac{1}{2}\left[(\sigma_x + \sigma_y) + \sqrt{(\sigma_x - \sigma_y)^2 + 4\tau_{xy}^2}\right]$$

$$\sigma_2 = \frac{1}{2}\left[(\sigma_x + \sigma_y) - \sqrt{(\sigma_x - \sigma_y)^2 + 4\tau_{xy}^2}\right]$$

At failure, the stresses in the two-dimensional complex system are

1. When $\sigma_2 \geq 0$

$$\sigma_{eq} = \sigma_1 = \sigma_{Yield} \ \text{(in tension)}$$

2. When $\sigma_1 \leq 0$

$$\sigma_{eq} = |\sigma_2| = \sigma_{Yield} \ \text{(in compression)} \qquad\qquad (9.1)$$

3. When $\sigma_1 > 0$, $\sigma_2 < 0$

$$\sigma_{eq} = \sigma_1 = \sigma_{Yield} \ \text{(in tension)}$$

and

$$\sigma_{eq} = |\sigma_2| = \sigma_{Yield} \ \text{(in compression)}$$

From the above comparison, it can be concluded that for an arbitrary state of stress

Failure (i.e. yielding) will occur when one of the principal stresses in a material is equal to the yield stress in the same material at failure in simple tension or compression.

The criterion was intended to work for brittle and ductile materials, while experimental evidence shows that it is approximately correct only for brittle materials. For brittle materials, both tensile and compressive strength should be checked since they are usually different.

9.2 MAXIMUM SHEAR STRESS CRITERION (TRESCA THEORY)

This criterion assumes that shear stress is the driving factor that causes failure of a material. According to this criterion, comparisons are made between the maximum shear stress of a material subjected to a complex stress system and that of the same material at failure when a simple tension is applied.

Two-dimensional complex state of stress

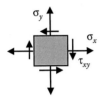

Simple tension *at failure*

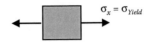

$\sigma_x = \sigma_{Yield}$

From Equation 7.4, the maximum shear stress is

Since there is no shear stress, $\sigma_x (= \sigma_{Yield})$ and $\sigma_y (= 0)$ are the two principal stresses. At the moment of failure, the maximum shear stress of the above state of stress is

$$\tau_{max} = \frac{1}{2}\sqrt{(\sigma_x - \sigma_y)^2 + 4\tau_{xy}^2}$$

$$\tau_{max} = \frac{\sigma_1 - \sigma_2}{2} = \frac{\sigma_x}{2} = \frac{\sigma_{Yield}}{2}$$

or

$$\tau_{max} = \frac{\sigma_1 - \sigma_2}{2}$$

At failure, the stresses in the two-dimensional complex system are

$$\sigma_{eq} = \sqrt{(\sigma_x - \sigma_y)^2 + 4\tau_{xy}^2} = \sigma_{Yield} \qquad (9.2)$$

or

$$\sigma_{eq} = \sigma_1 - \sigma_2 = \sigma_{Yield}$$

From the above comparison, it can be concluded that for an arbitrary state of stress

Failure (i.e. yielding) will occur when the maximum shear stress in the material is equal to the maximum shear stress in the same material at failure in simple tension.

9.3 DISTORTIONAL ENERGY DENSITY (VON MISES THEORY) CRITERION

This criterion assumes that distortional energy density (shear strain energy per unit volume) is the driving factor that causes failure of a material. According to this criterion, comparisons are made between the maximum shear strain energy per unit volume of a material subjected to a complex state of stress and that of the same material at failure when a simple tension is applied. Thus,

Two-dimensional complex stress system

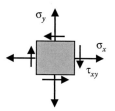

Simple tension *at failure*

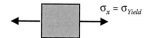

The shear strain energy per unit volume is

Since there is no shear stress, $\sigma_x(= \sigma_{Yield})$ and $\sigma_y(= 0)$ are the two principal stresses. At the moment of failure, the shear strain energy per unit volume is

$$U_s = \frac{1}{12G}[(\sigma_1 - \sigma_2)^2 + \sigma_2^2 + \sigma_1^2]$$

$$= \frac{1}{6G}[\sigma_1^2 + \sigma_2^2 - \sigma_1\sigma_2]$$

$$U_s = \frac{1}{6G}[\sigma_1^2 + \sigma_2^2 - \sigma_1\sigma_2]$$

$$= \frac{1}{6G}\sigma_{Yield}^2$$

or

$$U_s = \frac{1}{6G}[\sigma_x^2 - \sigma_x\sigma_y + \sigma_y^2 + 3\tau_{xy}^2]$$

At failure, the stresses in the two-dimensional complex system are

$$\sigma_{eq} = \sqrt{\sigma_1^2 + \sigma_2^2 - \sigma_1\sigma_2} = \sigma_{Yield} \tag{9.3}$$

$$\sigma_{eq} = \sqrt{\sigma_x^2 - \sigma_x\sigma_y + \sigma_y^2 + 3\tau_{xy}^2} = \sigma_{Yield}$$

From the above comparison, it can be concluded that for an arbitrary state of stress

---

Failure (i.e. yielding) will occur when the shear strain energy per unit volume in a material is equal to the equivalent value at failure of the same material in simple tension.

The application of the failure criterions depends on the modes of failure (e.g., failure by yielding or fracture). In general, the maximum principal stress criterion is valid for failure mode dominated by fracture in brittle materials, while Tresca and von Mises criterions are valid for general yielding mode of failure in ductile materials.

9.4 SPECIAL FORMS OF TRESCA AND VON MISES CRITERIONS

In many practical applications, at a surface point of a material there exists normal stress in one direction only. For example, at a surface point of the beam shown in Figure 9.3, the normal stress in the vertical direction is far smaller than that in the longitudinal direction and is usually ignored in the stress analysis of beams. If the longitudinal direction is defined

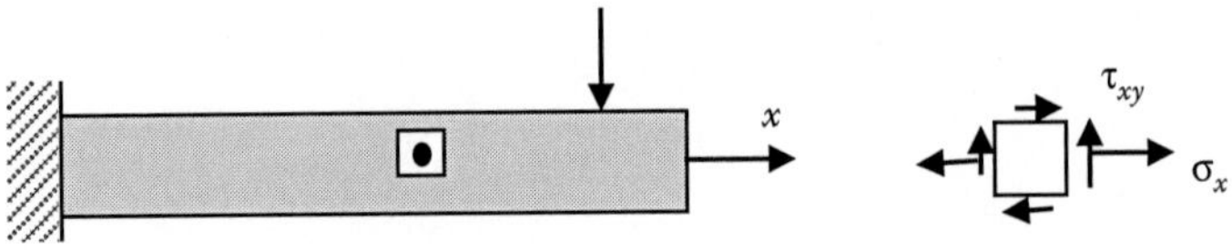

Figure 9.3 Beam subjected to bending and uniaxial tension.

as the x direction, $\sigma_y = 0$. Thus, Tresca and von Mises theories (Equations 9.2 and 9.3, respectively) are reduced to the following:

Tresca theory:

$$\sqrt{\sigma_x^2 + 4\tau_{xy}^2} = \sigma_{Yield} \qquad (9.4)$$

von Mises theory:

$$\sqrt{\sigma_x^2 + 3\tau_{xy}^2} = \sigma_{Yield} \qquad (9.5)$$

It is clear from Equations 9.4 and 9.5 that Tresca theory is more conservative than von Mises theory since the shear stress is factorised by 4 rather than 3.

9.5 KEY POINTS REVIEW

- A brittle material is likely to fail by fracture, and thus has higher compressive strength (holds greater compressive loads).
- A ductile material is likely to fail by yielding, that is, having permanent deformation.
- The maximum principal stress theory is best for brittle materials and can be unsafe for ductile materials.
- The maximum shear stress and the distortional energy density theories are suitable for ductile materials, while the former is more conservative than the latter.
- For brittle materials having a weaker tensile strength, reinforcement is usually required in the tension zone to increase the load-carrying capacity.

9.6 RECOMMENDED PROCEDURE OF SOLUTION

It is possible that a material may fail at any point within the material, but, in general, starts at a point where an equivalent stress defined above reaches a critical value first. Therefore, in a practical design, the application of the above criterions relies on identification of, for example, in the design of a beam, critical cross sections where maximum bending moment, twist moment or axial force may exist. On these critical sections, maximum normal and shear stresses are found. A recommended procedure of solution is shown in Figure 9.4.

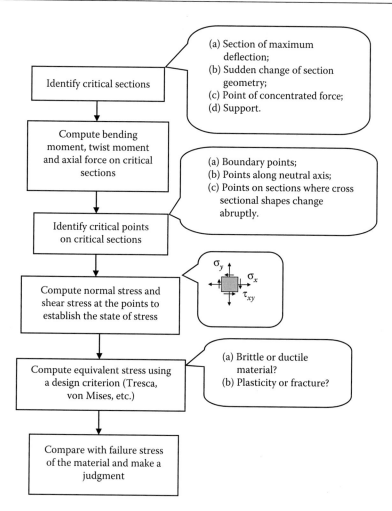

Figure 9.4 Flow chart for failure analysis.

9.7 EXAMPLES

EXAMPLE 9.1

Explain why concrete is normally reinforced with steel bars or rods when tensile forces are applied to a structure.

Solution

This question tests your understanding of the failure mode of brittle materials. This type of materials usually has different strength in tension and compression.

Concrete is a typical example of brittle materials that are weaker in tension, and have higher load capacity in compression. When concrete is subjected to tension, fracture is initiated at imperfections or micro-cracks, whereas the imperfections and micro-cracks are closed in compression and fracture is unlikely to occur. Therefore, steel or other types of reinforcements are needed in the tension zone to increase the tensile strength of the structure to prevent early fracture failure.

EXAMPLE 9.2

Codes of practice for the use of structural steel use either Tresca or von Mises criterion. For a beam member subjected to bending and shear, the criterions can be expressed as Tresca

$$\sqrt{\sigma_x^2 + 4\tau_{xy}^2} = \sigma_{Yield}$$

von Mises

$$\sqrt{\sigma_x^2 + 3\tau_{xy}^2} = \sigma_{Yield}$$

Verify the expression and state which criterion is more conservative.

Solution

The answer to this question is a direct application of Equations 9.2 and 9.3 to the state of stress where one of the normal stresses is zero.

For the beam shown in figure (a) below subjected to bending and shearing, the state of stress at the arbitrary point is represented by figure (b) below, where only one normal stress exists.

(a)   (b)

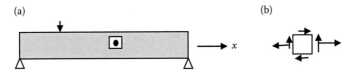

Assuming that the horizontal and vertical directions are, respectively, the x and y directions, $\sigma_y = 0$. Thus, the two principal stresses at the point are

$$\sigma_1 = \frac{1}{2}\left[\sigma_x + \sqrt{(\sigma_x)^2 + 4\tau_{xy}^2}\right]$$

$$\sigma_2 = \frac{1}{2}\left[\sigma_x - \sqrt{(\sigma_x)^2 + 4\tau_{xy}^2}\right]$$

Introducing the above principal stresses into Equations 9.2 and 9.3 yields, respectively, the special form of the two criterions.

Since $\sqrt{\sigma_x^2 + 4\tau_{xy}^2} > \sqrt{\sigma_x^2 + 3\tau_{xy}^2}$, for the same value of σ_{Yield}, a design by Tresca criterion demands smaller σ_x or/and τ_{xy}, that is, a reduction of applied loads or an increase of material usage. Thus, Tresca criterion is more conservative than von Mises criterion.

EXAMPLE 9.3

Consider a bar of cast iron under complex loading. The bar is subjected to a bending moment of $M = 39$ N m and a twist moment of $T = 225$ N m. The diameter of the bar is $D = 20$ mm. If the material of the bar fails at $\sigma_{Yield} = 128$ MPa in a simple tension test, will failure of the bar occur according to the maximum principal stress criterion?

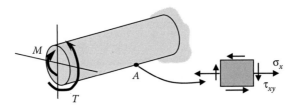

Solution

On any cross section of the beam, the maximum shear stress due to twisting is along the circumference and the maximum tensile stress due to bending is at the lowest edge. Thus, the state of stress at a point A taken from the lowest generator must be considered. The stresses at this point can be calculated from Equations 3.2, 5.1 and 5.2. The principal stresses can then be calculated from these stresses.

The shear stress due to torsion is

$$\tau_{xy} = \frac{Tr}{J} = \frac{TD/2}{\pi D^4/32} = \frac{32 \times 225\,\text{N m} \times 20 \times 10^{-2}\,\text{m}/2}{\pi \times (20 \times 10^{-3})^4}$$

$$= 143.24 \times 10^6\,\text{N/m}^2 = 143.24\,\text{MPa}$$

The normal stress due to bending is

$$\sigma_{xy} = \frac{My}{I} = \frac{MD/2}{\pi D^4/64} = \frac{64 \times 39\,\text{N m} \times 20 \times 10^{-3}\,\text{m}/2}{\pi \times (20 \times 10^{-3})^4}$$

$$= 49.66 \times 10^6\,\text{N/m}^2 = 49.66\,\text{MPa}$$

From Equation 7.2, the principal stresses at the point are

$$\sigma_1 = \frac{1}{2}\left[(\sigma_x + \sigma_y) + \sqrt{(\sigma_x - \sigma_y)^2 + 4\tau_{xy}^2}\right]$$

$$= \frac{1}{2}\left[(49.66\,\text{MPa} + 0) + \sqrt{(49.66\,\text{MPa} + 0)^2 + 4 \times (143.24\,\text{MPa})^2}\right]$$

$$= 169.96\,\text{MPa}$$

$$\sigma_2 = \frac{1}{2}\left[(\sigma_x + \sigma_y) - \sqrt{(\sigma_x - \sigma_y)^2 + 4\tau_{xy}^2}\right]$$

$$= \frac{1}{2}\left[(49.66\,\text{MPa} + 0) - \sqrt{(49.66\,\text{MPa} + 0)^2 + 4 \times (143.24\,\text{MPa})^2}\right]$$

$$= -120.3\,\text{MPa}$$

$$\sigma_1 > \sigma_{Yield}$$

The material has failed according to the maximum principal stress criterion.

EXAMPLE 9.4

In a tensile test on a metal specimen having a cross section of 20 mm × 10 mm, failure occurred at a load of 70 kN. A thin plate made from the same material is subjected to loads such that at a certain point in the plate the stresses are $\sigma_y = -70$ N/mm², $\tau_{xy} = 60$ N/mm². Determine from the von Mises and Tresca criterions the maximum allowable tensile stress, σ_x, that can be applied at the same point.

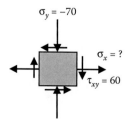

Solution

For the state of stress shown in figure above, the equivalent stresses of Equations 9.2 and 9.3 can be calculated and they are functions of σ_x. Comparing the equivalent stresses with the failure stress of the material at simple tension test yields the maximum allowable σ_x.

From the simple tension test

$$\sigma_{Yield} = \frac{70,000\,\text{N}}{20\,\text{mm} \times 10\,\text{mm}} = 350\,\text{N/mm}^2$$

Tresca criterion (Equation 9.2)

$$\sigma_{eq} = \sqrt{(\sigma_x - \sigma_y)^2 + 4\tau_{xy}^2} \leq \sigma_{Yield}$$
$$\sqrt{(\sigma_x - (-70))^2 + 4 \times 60^2} \leq 350$$

Therefore,

$$(\sigma_x + 70)^2 + 4 \times 60^2 \leq 350^2$$
$$\sigma_x^2 + 140x - 103,200 \leq 0$$

Thus,

$$\sigma_x \leq 259\,\text{N/mm}^2$$

von Mises criterion (Equation 9.3)

$$\sigma_{eq} = \sqrt{\sigma_x^2 - \sigma_x\sigma_y + \sigma_y^2 + 3\tau_{xy}^2} \leq \sigma_{Yield}$$
$$\sqrt{\sigma_x^2 - \sigma_x(-70) + (-70)^2 + 3 \times 60^2} \leq 350$$

Therefore,

$$\sigma_x^2 + 70\sigma_x - 106,800 \leq 0$$

Thus,

$$\sigma_x \leq 293.7\,\text{N/mm}^2$$

From the above solutions, it can be seen that the maximum allowable value of σ_x from Tresca criterion is smaller than that from von Mises criterion, and hence the resulting design is more conservative.

9.8 CONCEPTUAL QUESTIONS

1. Describe the differences between failure by fracture and failure by yielding.
2. Describe the failure mode of a cast iron bar subjected to a uniaxial tensile force. If the bar is made of mild steel, describe how the failure mode is different.
3. Describe the failure mode of a cast iron bar subjected to torsion. If the bar is made of mild steel, describe how the failure mode is different.

4. What is the definition of the maximum principal stress criterion? What type of material is it useful for?
5. What is the definition of the Tresca failure criterion? What type of material is it useful for?
6. What is the definition of the von Mises failure criterion? What type of material is it useful for?
7. Explain the term 'equivalent stress' as used in connection with failure criterions.
8. On what conditions can the simplified form of Tresca and von Mises criterions be used?

9.9 MINI TEST

PROBLEM 9.1

Explain why different failure criterions are needed in design and discuss what aspects should be considered in a typical design.

PROBLEM 9.2

In a ductile material there are four points at which the states of stress are, respectively, as shown in figure below.

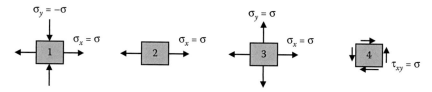

Which point fails first and which point fails at last? What conclusion can you draw from your analysis?

PROBLEM 9.3

The stresses at a point of a two-dimensional structural member are found as follows:

$$\sigma_x = 140\,\text{N/mm}^2, \quad \sigma_y = -70\,\text{N/mm}^2, \quad \tau_{xy} = 60\,\text{N/mm}^2$$

The material of the member has a yield stress in simple tension of 225 N/mm². Determine whether or not failure has occurred according to Tresca and von Mises criterions.

PROBLEM 9.4

On the beam section shown in figure below, there exists an axial force of 60 kN. Determine the maximum shear force that can be applied to the section using the Tresca and von Mises criterions. The material of the beam breaks down at a stress of 150 N/mm² in a simple tension test.

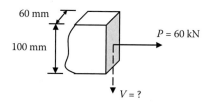

PROBLEM 9.5

The cantilever of circular cross section shown in figure below is made from steel, which when subjected to simple tension suffers elastic breakdown at a stress of 150 N/mm². If the cantilever supports a bending moment of 25 kN m and a torque of 50 kN m, determine the minimum diameter of the cantilever using the von Mises and Tresca criterions.

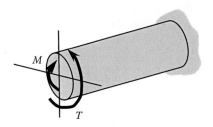

Chapter 10

Buckling of columns

Compression members, such as columns, are mainly subjected to axial compressive forces. The stress on a cross section in a compression member is therefore normal compressive stress. A short column usually fails due to yielding or shearing (Figure 10.1), depending on the material properties of the column.

However, when a compression member becomes longer, it could become laterally unstable and eventually collapse through sideways buckling at an axial compression. The compressive load could be far smaller than the one that would cause material failure of the same member. A simple test to illustrate this can be easily carried out by pushing the ends of a piece of spaghetti. A lateral deflection from the original position will be observed when the applied compression reaches a certain value, which is designated P_{cr} and called *critical buckling load* (Figure 10.2). The critical buckling load is the maximum load that can be applied to a column without causing instability. Any increase in the load will cause the column to fail by buckling.

The onset of the lateral deflection at the critical stage when $P = P_{cr}$ establishes the initiation of buckling, at which the structural system becomes unstable. P_{cr} represents the ultimate capacity of compression members, such as long columns and thin-walled structural components.

- Buckling may occur when a long or slender member is subjected to axial compression.
- A compression member may lose stability due to a compressive force that is smaller than the one causing material failure of the same member.
- The buckling failure of a compression member is related to the strength, stiffness of the material, and also the geometry (slenderness ratio) of the member.

For a long or slender member subjected to compression, considering material strength along is usually not sufficient in design.

10.1 EULER FORMULAS FOR COLUMNS

At the critical buckling load, a column may buckle or deflect in any lateral direction. In general, the flexural rigidity or stiffness of a column is not the same in all directions (Figure 10.3). By common sense, a column will buckle (deflect) in a direction related to the minimum flexural rigidity, EI, that is, the minimum second moment of area of the cross section.

The maximum axial load that a long, slender ideal column can carry without buckling was derived by Leonhard Euler in the eighteenth century, and termed *Euler formula* for columns. The formula was derived for ideal columns that are perfectly straight, homogeneous, and free from initial stresses.

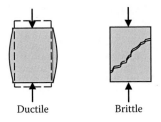

Figure 10.1 Strength failure of material subjected to compression.

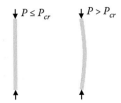

Figure 10.2 Buckling of long column.

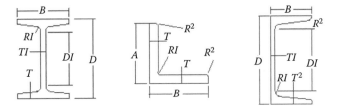

Figure 10.3 Commonly-used section profiles of structural column.

10.1.1 Euler formula for columns with pinned ends

The critical buckling load for the column shown in Figure 10.4 is

$$P_{cr} = \frac{\pi^2 EI}{L^2} \tag{10.1}$$

For the rectangular section shown, the flexural rigidity, EI, is related to the X–X axis, about which the second moment of area of the cross section is minimum. By Equation 10.1, a higher compressive force is required to cause buckling of a shorter column or a column with greater flexural rigidity.

Equation 10.1 is the Euler formula for a column with pinned ends and is often referred to as the fundamental case. In general, columns do not always have simply supported ends. Therefore, the formula for the critical buckling load needs to be extended to include other form of end supports.

10.1.2 Euler formulas for columns with other ends

Figure 10.5 shows comparisons between columns with different end supports. Comparing with the pin-ended column (case [a]), the buckling mode of the column with free–fixed ends (case [b]) is equivalent to a pinned column of doubled length.

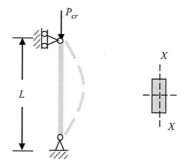

Figure 10.4 Buckling of column with pinned ends.

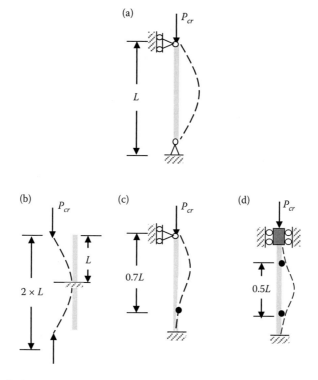

Figure 10.5 Buckling of columns with other ends.

Thus, for case (b)

$$P_{cr} = \frac{\pi^2 EI}{(2L)^2}$$

(10.2)

For cases (c) and (d), the Euler formulas can also be made by comparing their buckling modes with that of case (a). In cases (c) and (d), the deflection curves of the upper $0.7L$ and middle $0.5L$ show, respectively, a similar pattern to that shown in case (a). The considered deflection shapes are the segments between the two inflection points on the deflection curves. The distance between the two inflection points is called *effective length*, L_e, of a

Table 10.1 Effective length factor of compression members

Support at column ends	Effective length factor λ
Fixed–fixed	0.5
Pinned–fixed	0.7
Pinned–pinned	1
Free–free	1
Fixed–free	2

column when comparison is made with a similar column with two pinned ends. For a general case, therefore, the Euler formula can be written as

$$P_{cr} = \frac{\pi^2 EI}{L_e^2} = \frac{\pi^2 EI}{(\lambda L)^2} \tag{10.3}$$

where λ is the effective length factor that depends on the end restraints. Table 10.1 shows the values of λ for some typical end restraints.

10.2 LIMITATIONS OF EULER FORMULAS

The Euler formulas are applicable only while the material remains linearly elastic. To apply this limitation in practical designs, Equation 10.3 is rewritten as

$$P_{cr} = \frac{\pi^2 EI}{(\lambda L)^2} = \frac{\pi^2 EAr^2}{(\lambda L)^2} \frac{\pi^2 EA}{(\lambda L/r)^2} \tag{10.4}$$

or, the compressive critical stress is

$$\sigma_{cr} = \frac{P_{cr}}{A} = \frac{\pi^2 E}{(\lambda L/r)^2} \tag{10.5}$$

In Equations 10.4 and 10.5 the second moment of area, I, is replaced by Ar^2, where A is the cross-sectional area and r is its radius of gyration. $\lambda L/r$ is known as *slenderness ratio*. It is clear that

- If a column is long and slender, $\lambda L/r$ is large and σ_{cr} is small.
- If a column is short and has a large cross-sectional area, $\lambda L/r$ is small and σ_{cr} is large.

To ensure that the material of a column remains linearly elastic, the stress in the column before buckling must remain below the proportionality limit of the material, σ_p. Thus,

$$\sigma_{cr} = \frac{\pi^2 E}{(\lambda L/r)^2} \le \sigma_p \tag{10.6}$$

or

$$\frac{\lambda L}{r} \geq \pi \sqrt{\frac{E}{\sigma_p}} \qquad (10.7)$$

If the slenderness ratio of a column is smaller than the specified value in Equation 10.7, the Euler formulas are not applicable.

10.3 KEY POINTS REVIEW

- For a bar or a column subjected to compressive loading, a sudden failure (collapse) may occur. This type of failure is called *buckling*.
- A compression member may fail due to buckling at a stress level, which is far below the compressive strength of the material.
- The minimum compressive load that causes buckling of a compressive member is known as critical buckling load.
- For a compression member, the critical buckling load is proportional to its flexural rigidity and inversely proportional to the square of its length.
- The end support conditions have a significant influence on the critical load, which determine the number of inflection points on the deflected column.
- The closer the inflection points of a column are, the higher the critical buckling load of the column is.
- The Euler formula for buckling of a column is based on the following assumptions:
  - The column is initially perfectly straight.
  - The compression is applied axially.
  - The column is very long in comparison with its cross-sectional dimensions.
  - The column is uniform throughout and the proportional limit of materials is not exceeded.
- The ratio of the effective length to the radius of gyration of a column is defined as *slenderness ratio*.
- Slenderness ratio governs the critical buckling load: the larger the slenderness ratio is, the lesser the strength of a column. This means that the buckling resistance decreases as the slenderness ratio increases.
- The application of Euler formulas depends on elasticity rather than compressive strength of material.
- For a fixed column cross section, length and end supports, the critical buckling load capacity depends only upon Young's modulus E. Since there is little variation in E among different grades of steel, there is no advantage in using a high-strength steel.

10.4 EXAMPLES

EXAMPLE 10.1

Derive the Euler formula for the column with pinned ends shown in figure below.

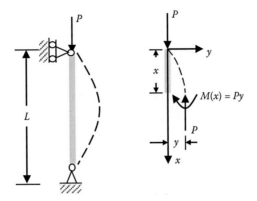

Solution

The buckled shape shown in figure above is possible only when the applied load is greater than the critical load. Otherwise, the column remains straight. The solution of this problem is to seek the relationship between the applied axial load and the lateral of the column.

From Equation 6.1, the deflection and the bending moment on the cross section at an arbitrary location x is related by

$$\frac{d^2y}{dx^2} = -\frac{M(x)}{EI} = -\frac{Py}{EI}$$

The above equation can be rewritten as

$$\frac{d^2y}{dx^2} + k^2y = 0$$

where $k^2 = P/EI$.

The equation is a linear ordinary differential equation of second order and its solution is

$$y = A\sin kx + B\cos kx$$

where A and B are arbitrary constants that can be determined from the end support conditions. These conditions are

at $x = 0$; $y(0) = 0$
at $x = L$; $y(L) = 0$

Hence,

$y(0) = 0$; $A\sin 0 + B\cos 0 = 0$

or

$B = 0$
$y(L) = 0$; $A\sin kL + B\cos kL = 0$

or

$$A \sin kL = 0$$

The above condition can be satisfied by letting $A = 0$, but this will lead to no lateral deflection or buckling. Alternatively, the condition can be satisfied by taking

$$kL = m\pi$$

where m is a nonzero integer. Thus,

$$\sqrt{\frac{P}{EI}} L = m\pi$$

or

$$P = \frac{m^2 \pi^2 EI}{L^2}$$

The above solution provides a series of buckling loads at which the column will buckle with different shapes. Obviously, the lowest value of the force is the critical buckling load as defined in Figure 10.2, and takes the value when $m = 1$, that is,

$$P_{cr} = \frac{\pi^2 EI}{L^2}$$

This result is identical to the solution given in Equation 10.1.

EXAMPLE 10.2

The pin-connected plane steel truss shown in figure below carries a concentrated vertical force F. Assuming that both members have a circular section with a diameter of 80 mm, determine the critical buckling load of the truss ($E = 200$ GPa).

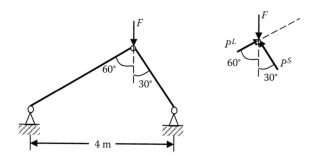

Solution

The critical load of the truss is the minimum load that will cause buckling of at least one member. The smallest critical buckling load of the two pin-ended members must be calculated first, to which the critical buckling load of the truss is related through equilibrium of the joint.

For the long member

$$P_{cr}^L = \frac{\pi^2 EI}{L^2} = \frac{\pi^2 \times 20 \times 10^9 \, \text{Pa} \times \pi \times (80 \times 10^{-3})^4/64}{(4\,\text{m} \times \cos 30°)^2} = 330.7\,\text{kN}$$

For the short member

$$P_{cr}^S = \frac{\pi^2 EI}{L^2} = \frac{\pi^2 \times 20 \times 10^9 \, \text{Pa} \times \pi \times (80 \times 10^{-3})^4/64}{(4\,\text{m} \times \sin 30°)^2} = 992.2\,\text{kN}$$

Apparently, the long member will buckle first at an axial compression of 330.7 kN. For the equilibrium of the pin joint, taking $p^L = P_{cr}^L$ and resolving all the forces in the P^L direction yield:

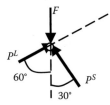

$$P_{cr}^L = F_{cr} \times \cos 60°$$

or

$$F_{cr} = \frac{P_{cr}^L}{\cos 60°} = \frac{330.7\,\text{kN}}{\cos 60°} = 661.4\,\text{kN}$$

When $F = F_{cr} = 661.4$ kN, the axial compression in the short member is

$$P^S = F \times \cos 30° = 661.4 \times \frac{\sqrt{3}}{2} = 572.8\,\text{kN}$$

The short member is stable. Thus, the critical buckling load of the truss is 661.4 kN.

EXAMPLE 10.3

A beam–column structure (figure below) consists of a beam of rectangular section (200 mm × 300 mm) and a column of I-shaped section (UBS 178 × 102 × 19). The allowable stress of the beam is $\sigma_{allow} = 100$ MPa. The column is made of steel having $E = 200$ GPa and $\sigma_p = 200$ MPa. Determine the maximum point load that can be applied along the beam.

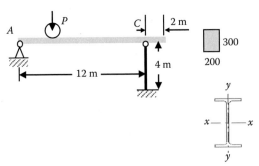

Solution

To determine the maximum load, both the beam and the column should be considered. The bending strength of the beam and the instability of the column must be checked. For the beam, when the force is applied at the mid-span, the maximum bending moment and the maximum normal stress occur on the mid-span section. The compression in the column is maximum when the force moves to the free end of the beam.

For the beam, when the force is applied at the mid-span, the maximum bending moment occurs on the mid-span cross section and is

$$M_{max} = \frac{PL}{4} = \frac{P \times 12\,m}{4} = 3P(N\,m)$$

The maximum normal stress on the section is

$$\sigma_{max} = \frac{M_{max}h/2}{bh^3/12} = \frac{3P \times 300 \times 10^{-3}\,m/2}{200 \times 10^{-3}\,m \times (300 \times 10^{-3}\,m)^3/12} = P \times 10^3\,N/m^2$$

From $\sigma_{max} \leq \sigma_{allow} = 100$ MPa

$$P \leq 100\,kN$$

For the column (UBS $178 \times 102 \times 19$), the radius of gyration can be found as

$$r_x = 7.48\,cm, \quad r_y = 2.37\,cm, \quad A = 24.3\,cm^2$$

The column may buckle in the plane of X–X or Y–Y, depending on the slenderness ratios relative to these directions. In the X–X plane, the column can be taken as supported by pin and fixed ends and in the Y–Y plane the column can be considered as one with free–fixed ends. Thus, the slenderness ratios of the column about the X–X and Y–Y axes are, respectively (see Table 10.1),

$$\frac{\lambda_x L}{r_x} = \frac{2 \times 4\,m}{7.48 \times 10^{-2}\,m} = 107$$

$$\frac{\lambda_y L}{r_y} = \frac{0.7 \times 4\,m}{2.37 \times 10^{-2}\,m} = 118.1$$

From Equation 10.7

$$\frac{\lambda_x L}{r_x} = 107 > \pi\sqrt{\frac{E}{\sigma_P}} = \pi\sqrt{\frac{200 \times 10^9\,Pa}{200 \times 10^6\,Pa}} = 99.3$$

$$\frac{\lambda_y L}{r_y} = 118.1 > \pi\sqrt{\frac{E}{\sigma_P}} = \pi\sqrt{\frac{200 \times 10^9\,Pa}{200 \times 10^6\,Pa}} = 99.3$$

Since the slenderness ratios of the column are larger than the value specified in Equation 10.3, the Euler formulas are valid. Buckling about the Y–Y axis is more likely due to the greater slenderness ratio. Thus, from Equation 10.4, the critical buckling load of the column is

$$P_{cr} = \frac{\pi^2 EA}{(\lambda_y L/r_y)^2} = \frac{\pi^2 \times 200 \times 10^9 \, \text{Pa} \times 24.3 \times 10^{-4} \, \text{m}^2}{(118.1)^2} = 343.9 \, \text{kN}$$

The maximum compression in the column caused by the applied force occurs when the force acts at the free end of the beam. The compression can be easily calculated from the equilibrium of the beam by taking moment about A.

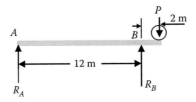

$$12 \, \text{m} \times R_B - (12 + 2)\text{m} \times P = 0$$

which yields

$$R_B = \frac{7}{6} P$$

or

$$P = \frac{6}{7} R_B$$

The compression in the column is numerically equal to R_B. When R_B reaches the critical value, that is, $R_B = P_{cr}$

$$P \leq \frac{6}{7} R_B = \frac{6}{7} P_{cr} = \frac{6}{7} \times 343.9 \, \text{kN} = 294.8 \, \text{kN}$$

Compared with the maximum force obtained from applying the strength condition of the beam, which is less than 294.8 kN, the maximum applied force that the structure can carry is therefore 100 kN.

EXAMPLE 10.4

A steel column of length 12 m (figure below) is pinned at both ends and has an American Standard Steel Channel section (C130 × 10). Young's modulus of the steel is $E = 200$ GPa, and the elastic limit of the material is 200 MPa.

a. Determine the critical buckling load of the column.
b. If a column is built up of two C130 × 10 channels placed back to back at a distance of d and connected to each other at an interval of h along the length of the column, determine the values of d and h and calculate the critical buckling load of the column.

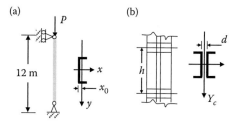

(a)   (b)

Solution

A single channel section is not symmetric. Such a column will buckle in the direction of minimum radius of gyration. When the two channel sections are combined together (figure [b] above), the radius of gyration about the x-axis is a constant, while the radius of gyration about the Y_c direction depends on the back-to-back distance d. The best design for this combined section is to achieve the same radius of gyration about both the x and Y_c directions, from which distance d can be determined. Between the two connection points, the best design is to make sure that the two channels do not buckle individually and behave as a composite unit before the critical buckling load of the full-length column has been reached. This can be used as the requirement to determine h.

The section properties of C130 × 10 are as follows (figure [a] above):

$$A = 12.71\,\text{cm}^2, I_x = 312\,\text{cm}^4, I_y = 19.9\,\text{cm}^4$$
$$r_x = 4.95\,\text{cm}, r_y = 1.25\,\text{cm}, x_0 = 1.23\,\text{cm}$$

a. For the single section, since $r_y < r_x$ and

$$\frac{\lambda_y L}{r_y} = \frac{1 \times 12 \times 10^2\,\text{cm}}{1.25\,\text{cm}} = 960$$

$$\geq \pi\sqrt{\frac{E}{\sigma_e}} = \pi\sqrt{\frac{200 \times 10^9\,\text{Pa}}{200 \times 10^6\,\text{Pa}}} = 99.35$$

the column will fail due to buckling at

$$P_{cr} = \frac{\pi^2 E I_y}{(\lambda_y L)^2} = \frac{\pi^2 \times 200 \times 10^9\,\text{Pa} \times 19.9 \times 10^{-8}\,\text{m}^4}{(1 \times 12\,\text{m})^2} = 2.73\,\text{kN}$$

b. For the combined section, the second moment of area about the x-axis is twice that of the single section. The second moment of area about the Y_c-axis varies depending on the back-to-back distance d. For a single section, the second moment of area about the Y_c-axis (figure [b] above) is calculated by the parallel axis theorem (Equation 5.4).

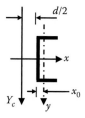

$$I_{Y_c} = I_y + A\left(x_0 + \frac{d}{2}\right)^2 = 19.9\,\text{cm}^4 + 12.71\,\text{cm}^2 \times \left(1.23\,\text{cm} + \frac{d}{2}\right)^2$$

To achieve an equal radius of gyration about both the x- and the Y_c-axis requires

$$2 \times I_{Y_c} = 2 \times I_x \quad \text{or}$$

$$19.9\,\text{cm}^4 + 12.71\,\text{cm}^2 \times \left(1.23\,\text{cm} + \frac{d}{2}\right)^2 = 312\,\text{cm}^4$$

Thus,

$$d = 2\left(\sqrt{\frac{312 - 19.9}{12.71}} - 1.23\right) = 7.13\,\text{cm}$$

Within the two connection points of distance h, buckling of individual columns occurs when the compression in the columns reaches a certain level. To prevent this local failure, the minimum requirement is that the critical local buckling load (local buckling mode) is greater than or at least equal to half of the critical buckling load of the entire column (overall buckling mode). Thus, the slenderness ratio of a single column between the two connection points must be smaller or at least equal to two times the slenderness ratio of the entire column having the combined section:

$$\frac{\lambda_y h}{r_y} \le 2 \times \frac{\lambda_y L}{2r_{Y_c}} = 2 \times \frac{\lambda_x L}{2r_x} = \frac{\lambda_x L}{r_x}$$

Hence,

$$\frac{1 \times h}{1.25 \times 10^{-2}\,\text{m}} \le \frac{1 \times 12\,\text{m}}{4.95 \times 10^{-2}\,\text{m}}$$

From which

$$h < 3.03\,\text{m}$$

With the above-calculated h and d, the slenderness ratio of the composite column for the overall buckling satisfies

$$\frac{\lambda_x L}{2r_x} = \frac{1 \times 12\,\text{m}}{2 \times 4.95 \times 10^{-2}\,\text{m}} = 121.2 \ge \pi\sqrt{\frac{E}{\sigma_p}}$$

The composite column will fail by buckling, and the critical buckling load for the composite column with the above-calculated d and h is

$$P_{cr} = \frac{\pi^2 E(2 \times I_x)}{(\lambda_x L)^2} = \frac{\pi^2 \times 200 \times 10^9\,\text{Pa} \times 2 \times 312 \times 10^{-8}\,\text{m}^4}{(1 \times 12\,\text{m})^2} = 85.54\,\text{kN}$$

By constructing the composite section, the critical buckling load of the column increases from 2.73 to 85.54 kN, an increase of 31 times.

EXAMPLE 10.5

Derive the solution for buckling of columns with general end conditions.

Solution

Euler formula (Equation 10.3) is given on the basis of knowing the effective length of a column. The effective length factors for selected cases are given in Table 10.1. This example shows how these factors can be calculated and how the buckling load of columns with other support conditions can be evaluated. The solution starts with solving general bending equation from Example 10.1. In order to include the solution for columns with general end conditions (each end should have two conditions to describe), a fourth-order differential equation is needed.

From Example 10.1, the deflection and bending moment for a buckled column is related by

$$\frac{d^2y}{dx^2} + k^2y = 0$$

Differentiating the equation twice yields

$$\frac{d^4y}{dx^4} + k^2\frac{d^2y}{dy^2} = 0$$

where $k^2 = P/EI$. The solution of this fourth-order differential equation is

$$y = C_1 \sin kx + C_2 \cos kx + C_3 x + C_4$$

The four unknown constants are to be determined through the introduction of support conditions. The derivatives of the above solution are given as

$$\frac{dy}{dx} = C_1 k \cos kx - C_2 k \sin kx + C_3$$

$$\frac{d^2y}{dx^2} = C_1 k^2 \sin kx - C_2 k^2 \cos kx$$

$$\frac{d^3y}{dx^3} = -C_1 k^3 \cos kx - C_2 k^2 \sin kx$$

The above derivatives are related to the rotation, bending moment and shear force (Equations 6.1 and 6.2) of the column. Introducing appropriate deflection, rotation, bending moment and shear force conditions at both ends of the column leads to a solution of k, from which the buckling loads can be calculated.

For a column with fixed ends (figure below), for example, the end conditions are

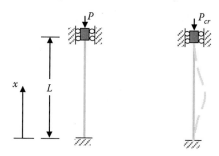

at $x = 0$, $y(0) = \dfrac{dy}{dx}(0) = 0$

at $x = L$, $y(L) = \dfrac{dy}{dx}(L) = 0$

Introducing the above conditions into the solutions yields

$y(0) = 0 : C_2 + C_4 = 0$

$\dfrac{dy}{dx}(0) = 0 : C_1 k + C_3 = 0$

$y(L) = 0 : C_1 \sin kL + C_2 \cos kL + C_3 L + C_4 = 0$

$\dfrac{dy}{dx}(L) = 0 : C_1 k \cos kL - C_2 k \sin kL + C_3 = 0$

To satisfy this set of equations, C_1, C_2, C_3 and C_4 could take zero, which are trivial solutions that show no deflection of the column. However, when the column buckles, at least one of the four constants must not be zero. This requires

$$\det \begin{vmatrix} 0 & 1 & 0 & 1 \\ k & 0 & 1 & 0 \\ \sin kL & \cos kL & L & 1 \\ k \cos kL & -k \sin kL & 1 & 0 \end{vmatrix} = 0$$

The evaluation of this determinant yields

$2 \cos kL + kL \sin kL - 2 = 0$

The minimum value of kL satisfying the above equation is $kL = 2\pi$. Thus,

$$\sqrt{\dfrac{P}{EI}} L = 2\pi$$

or $P = \dfrac{4\pi^2 EI}{L^2} = \dfrac{\pi^2 EI}{(0.5L)^2}$

The effective length factor is 0.5, which is exactly the same as the one shown in Table 10.1.

10.5 CONCEPTUAL QUESTIONS

1. Define the terms 'column' and 'slenderness' and explain the term 'slenderness ratio' as applied to columns.
2. Explain how the Euler formula for pin-ended columns can be modified for columns having one or both ends fixed.
3. Which of the following statement defines the term 'critical buckling load'?
   a. The stress on the cross section due to the critical load equals the proportional limit of the material.
   b. The stress on the cross section due to the critical load equals the strength of the material.

   c. The maximum compressive axial force that can be applied to a column before any lateral deflection occurs.

   d. The minimum compressive axial force that can be applied to a column before any lateral deflection occurs.

   e. The axial compressive axial force that causes material failure of a column.

4. Why does the Euler formula become unsuitable at certain values of slenderness ratio?

5. Define the term 'local buckling'.

6. A column with the following cross sections (figure below) is under axial compression; in which direction may the column buckle?

7. If two columns made of the same material have the same length, cross-sectional area and end supports, are the critical buckling loads of the two columns the same?

8. The long bar shown in figure below is fixed at one end and elastically restrained at the other. Which of the following evaluations of its effective length factor is correct?

   (a) $\lambda < 0.5$ (b) $0.5 < \lambda < 0.7$ (c) $0.7 < \lambda < 2$ (d) $\lambda > 2$

9. Two compression members are made of the same material and have equal cross-sectional area. Both members also have the same end supports. If the members have square and circular sections, respectively, which of the member has greater critical buckling load?

10. Consider the columns shown in figure below. All columns are made of the same material and have the same slenderness ratio. The columns are all pinned at both ends, two of which also have intermediate supports (bracing). If the critical buckling load of column (a) is P_{cr}, calculate the critical buckling loads of the other two columns.

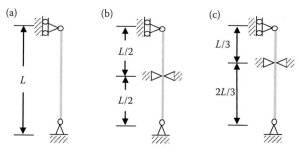

10.6 MINI TEST

PROBLEM 10.1

Describe the factors that affect critical buckling load of a column and how stability capacity of the column can be enhanced.

PROBLEM 10.2

A compressive axial force is applied to two steel rods that have the same material, support conditions and cross-sectional area. If the cross sections of the two rods are, respectively, solid circular and hollow circular, which rod will buckle at a lower force and why?

PROBLEM 10.3

A bar of solid circular cross section has built-in end conditions at both ends and a cross-sectional area of 2 cm². The bar is just sufficient to support an axial load of 20 kN before buckling. If one end of the bar is now set free from any constraint, it is obvious that the bar has a lower critical buckling load (figure below). The load-carrying capacity, however, can be maintained without consuming extra material but by making the bar hollow to increase the second moment of cross-sectional area. Determine the external diameter of such a hollow bar.

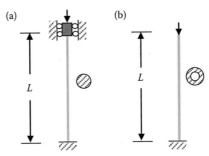

PROBLEM 10.4

The pin-joined frame shown in figure below carries a downward load P at C. Assuming that buckling can only occur in the plane of the frame, determine the value of P that will cause instability. Both members have a square section of 50 mm × 50 mm. Take $E = 70$ GPa for the material.

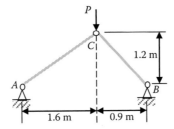

PROBLEM 10.5

A column is built up of two channels with two thin plates bolted to the flanges as shown in figure below. Calculate the required distance between the backs of the two channels in order to achieve an equal buckling resistance about both the X- and the Y-axis. If the column is 1 m long and fixed at both ends, determine its critical buckling load. Take $E = 80$ GPa.

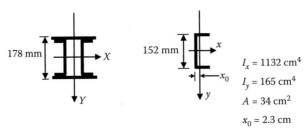

Chapter 11

Energy method

In the previous chapters, the structural and stress problems were solved on the basis of static equilibrium by considering the relationship between the internal forces (stresses) and the externally applied loads. The same problems can also be solved on the basis of the principle of conservation of energy by considering the energy built up within a body. The energy is stored due to the deformation in relation to the work done by the externally applied loads. In general, the principle of conservation of energy in structural and stress analysis establishes the relationships between stresses, strains or deformations, material properties and external loadings in the form of energy or work done by internal and external forces. The basic concepts of work and energy are as follows:

- Work is defined as the product of a force and the distance in the direction of the force.
- Energy is defined as the capacity to do work.

Unlike stresses, strains or displacement, energy or work is a scalar quantity. Simple application of an energy method is to equalise the work done by external loads and the energy stored in a deformed body, while the most powerful method that can effectively solve a complex structural and stress problem is based on the principle of virtual work.

11.1 WORK AND STRAIN ENERGY

11.1.1 Work done by a force

In order to accomplish work on an object, there must be a force exerted on the object and it must move in the direction of the force. The unit of work is, for example, N m, and is also called Joules.

In Figure 11.1, the work done by the force, F, to move the mass by a horizontal distance, d, is

$$Work = F \times \cos\theta \times d \tag{11.1}$$

11.1.2 Strain energy

In a solid deformable body, the work done by stresses on their associated deformation (strains) is defined as strain energy. In general, strain energy is computed as

$$u = \int_0^\varepsilon \sigma d\varepsilon \tag{11.2}$$

203

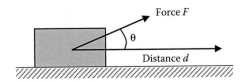

Figure 11.1 Work done by a force.

Table 11.1 Strain energy

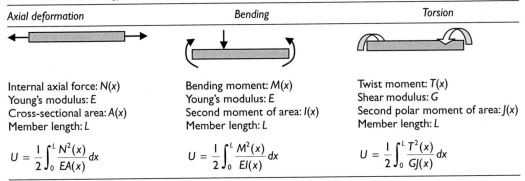

Axial deformation	Bending	Torsion
Internal axial force: $N(x)$ Young's modulus: E Cross-sectional area: $A(x)$ Member length: L	Bending moment: $M(x)$ Young's modulus: E Second moment of area: $I(x)$ Member length: L	Twist moment: $T(x)$ Shear modulus: G Second polar moment of area: $J(x)$ Member length: L
$U = \dfrac{1}{2}\displaystyle\int_0^L \dfrac{N^2(x)}{EA(x)}dx$	$U = \dfrac{1}{2}\displaystyle\int_0^L \dfrac{M^2(x)}{EI(x)}dx$	$U = \dfrac{1}{2}\displaystyle\int_0^L \dfrac{T^2(x)}{GJ(x)}dx$

and

$$U = \int_{vol} u\,dV \tag{11.3}$$

where u denotes strain energy per unit volume and U is the total strain energy stored in a body. The strain energies due to different types of deformation are listed in Table 11.1.

If a member is subjected to a combined action of axial force, bending moment and torque, the strain energy is

$$U = \frac{1}{2}\int_0^L \frac{N^2(x)}{EA}dx + \frac{1}{2}\int_0^L \frac{M^2(x)}{EI}dx + \frac{1}{2}\int_0^L \frac{T^2(x)}{GJ}dx \tag{11.4}$$

11.2 SOLUTIONS BASED ON ENERGY METHOD

The linearly elastic system shown in Figure 11.2 is subjected to a set of point loads. The following theorems establish the relationship between the applied forces and the displacements in the directions of the forces.

11.2.1 Castigliano's first theorem

If the strain energy of an elastic structural system is expressed in terms of n independent displacements, $\delta_1, \delta_2, ..., \delta_n$, associated with a system of prescribed forces, $F_1, F_2, ..., F_n$, the first

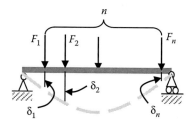

Figure 11.2 Strain energy due to the work done by the point forces.

partial derivative of the energy with respect to any of these displacements, δ_i, is equal to force, F_i, at point i in the direction of δ_i.

The mathematical expression of the theorem is

$$\frac{\partial U}{\partial \delta_i} = F_i \qquad (11.5)$$

In order to use Castigliano's first theorem, the strain energy must be expressed in terms of displacements $\delta_i \ (i = 1, 2, ..., n)$.

11.2.2 Castigliano's second theorem

If the strain energy of a linear elastic structural system is expressed in terms of n independent forces, $F_1, F_2, ..., F_n$, associated with a system of displacements, $\delta_1, \delta_2, ..., \delta_n$, the first partial derivative of the energy with respect to any of these forces, F_i, is equal to displacement, δ_i, at point i in the direction of F_i.

The mathematical expression of the theorem is

$$\frac{\partial U}{\partial F_i} = \delta_i \qquad (11.6)$$

In order to use Castigliano's second theorem, the strain energy must be expressed in terms of forces $F_i \ (i = 1, 2, ..., n)$.

For the energy expressed in the form of Equation 11.4

$$\delta_i = \frac{\partial U}{\partial F_i} = \int_0^L \frac{N(x)}{EA} \frac{\partial N(x)}{\partial F_i} dx + \int_0^L \frac{M(x)}{EI} \frac{\partial M(x)}{\partial F_i} dx + \int_0^L \frac{T(x)}{GJ} \frac{\partial N(x)}{\partial F_i} dx \qquad (11.7)$$

In the two theorems (Equations 11.5 and 11.6), the general force, F_i, can be a moment for which the associated displacement is the rotation at the same point.

11.3 VIRTUAL WORK AND THE PRINCIPLE OF VIRTUAL WORK

11.3.1 Virtual work

A force, F, which may be real or imaginary and acts on an object that is in equilibrium under a given system of loads, is said to do virtual work when the object is imagined to undergo a real or imaginary displacement in the direction of the force. Since the force and/or the displacements are not necessarily real, the work done is called *virtual work*.

Virtual work is classified as follows:

- External virtual work if the work is done by real or imaginary externally applied forces on an unrelated real or imaginary displacements of a system.
- Internal virtual work if the work is done by real or imaginary stresses on unrelated real or imaginary strains of a system.

To accomplish virtual work, the virtual displacement or deformation can be any unrelated deformation, but must satisfy the support conditions or boundary constraints of the system. For example, Figure 11.3 shows a beam under two separate load–displacement systems. Case (a) shows the beam's real deformation under the action of the loads shown. Case (b) shows the same beam that undergoes a real deformation under the action of a force, F_b.

If the deformation of case (a) is taken as the virtual deformation of case (b), additional to the real deformation case (b) has already had, the external virtual work done by F_b is

$$\delta W_e = F_b \times d_a$$

The internal virtual work done by the stresses caused by F_b in case (b) on the virtual strains (real strain of case [a]) is

$$\delta W_i = \int_L \frac{M_b(x)M_a(x)}{EI} \, dx$$

where $M_a(x)$ and $M_b(x)$ are, respectively, the bending moments in case (a) and case (b).

11.3.2 The principle of virtual work

The principle of virtual work is one of the most effective methods for calculating deflections (deformation). The principle of virtual work states as follows: if an elastic body under a system of external forces is given a small virtual displacement, then the increase in work done by the external forces is equal to the increase in strain energy stored.

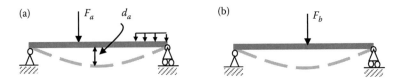

Figure 11.3 Illustration of virtual work.

In Figure 11.3, by taking the deformation of case (a) as the virtual deformation of case (b), the principle of virtual work yields

$$\delta W_e = \delta W_i$$

or

$$F_b \times d_a = \int_L \frac{M_a(x)M_b(x)}{EI}\, dx \tag{11.8}$$

In particular, if F_b is chosen as a unit force, the above equation yields the mid-span deflection of the beam under the loads shown in case (a), that is,

$$d_a = \int_L \frac{M_a(x)M_b(x)}{EI}\, dx \tag{11.9}$$

Equation 11.9 is called solution of the *unit load method*. For a general case involving axial deformation, bending and torsion, the solution is

$$\Delta_P = \int_0^L \frac{N_a(x)N_b(x)}{EA}\, dx + \int_0^L \frac{M_a(x)M_b(x)}{EI}\, dx + \int_0^L \frac{T_a(x)T_b(x)}{GJ}\, dx \tag{11.10}$$

where
  $N_a(x)$, $M_a(x)$ and $T_a(x)$ are the real internal forces of a system subject to externally applied loads.
  $N_b(x)$, $M_b(x)$ and $T_b(x)$ are the internal forces if the same system is subject to a unit force (moment) at a particular point and in a particular direction.
  Δ_p is the displacement/rotation of the system at the location, subject to the real external loads, where the unit force (moment) is applied.

Note: The displacement/rotation is in the same direction of the applied unit force/moment if the computed displacement/rotation is positive. Otherwise, it is in the opposite direction of the applied unit load.

11.3.3 Deflection of a truss system

Within the members of a truss system (pin-jointed frame), there is no bending and twist moments and the axial forces are constant along the length of each member. Thus, for a truss comprising n members, Equation 11.10 is reduced to

$$\Delta_P = \int_0^L \frac{N_a(x)N_b(x)}{EA}\, dx + \sum_{i=1}^{n} \frac{N_a^{(i)}N_b^{(i)}}{E_i A_i} L_i \tag{11.11}$$

where $N_a^{(i)}$ and $N_b^{(i)}$, E_i, A_i and L_i are the respective forces, Young's modulus, cross-sectional area and length of the ith member. They are usually all constant within a member.

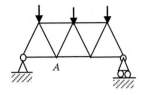

Figure 11.4 Deflection of truss.

Application of the unit load method equation (11.11) to find deflection of a truss system, for example the vertical deflection at joint *A* of the truss shown in Figure 11.4, follows the procedure below:

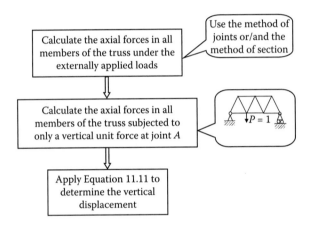

11.4 KEY POINTS REVIEW

- Any force will do work if it is associated with a deformation (displacement).
- Work is a scalar quality.
- Energy can be stored due to tension, compression, shearing, bending and twisting.
- Strain energy can be related to the change in dimension of body.
- For a material subject to externally applied loads, the work done by the applied loads must equal the strain energy stored in the material.
- Virtual work can be done by real force on imaginary (virtual) displacement or imaginary (virtual) force on real displacement.
- An imaginary (virtual) deformation of a system can be any possible deformation of the system that satisfies the support conditions, including the real deformation of the system.
- Castigliano's theorems provide relationships between a particular deformation and a particular force at a point.
- The unit load method provides a convenient tool for computing displacements in structural and stress analysis. It is applicable for both linear and nonlinear materials.
- A displacement obtained from applying the unit load method can be negative. In such a case, the displacement is in the opposite direction of the applied unit force.

11.5 EXAMPLES

EXAMPLE 11.1

The indeterminate truss shown in figure below consists of three members having the same value of EA. The truss is subjected to a downward force F as shown. Determine the axial forces in the three members by using Castigliano's first theorem.

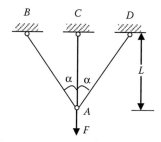

Solution

In order to use Castigliano's first theorem, the strain energy must be expressed in terms of displacement. For such a symmetric system, we assume that joint A has a small vertical displacement Δ, which is the elongation of member AC. The elongation of members AB and AD can be easily determined from geometrical relations. The elongations are used to calculate the strain energy that can be subsequently used, along with the equilibrium condition, to compute the axial forces within each of the members.

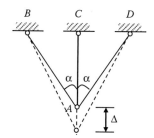

Assume that the elongation of AC, AB and AD are, respectively, Δ, Δ_{AB} and Δ_{AD}. For element AC under axial tension, applying Hooke's law yields

$$\sigma = Es$$

or

$$\frac{N_{AC}}{A} = E\frac{\Delta}{L}$$

then

$$N_{AC} = EA\frac{\Delta}{L}$$

The strain energy of this member is (Table 11.1)

$$U_{AC} = \frac{1}{2}\int_0^L \frac{N^2(x)}{EA}\,dx = \frac{1}{2}EA\left(\frac{\Delta}{L}\right)^2 \times L = \frac{EA\Delta^2}{2L}$$

For members AB and AD, since their length is $L/\cos\alpha$ and elongations are, respectively, Δ_{AB} and $\Delta_{AD,}$ the strain energies stored in these two members are

$$U_{AB} = \frac{EA(\Delta_{AB})^2}{2(L/\cos\alpha)} = \frac{EA(\Delta_{AB})^2 \cos\alpha}{2L}$$

$$U_{AD} = \frac{EA(\Delta_{AD})^2}{2(L/\cos\alpha)} = \frac{EA(\Delta_{AD})^2 \cos\alpha}{2L}$$

The total strain energy stored in the truss system is

$$U = U_{AB} + U_{AC} + U_{AD}$$
$$= \frac{EA(\Delta_{AB})^2 \cos\alpha}{2L} + \frac{EA\Delta^2}{2L} + \frac{EA(\Delta_{AD})^2 \cos\alpha}{2L}$$

By Castigliano's first theorem, the axial forces in these members are, respectively,

$$N_{AC} = \frac{\partial U}{\partial \Delta} = \frac{EA\Delta}{L}$$

$$N_{AB} = \frac{\partial U}{\partial \Delta_{AB}} = \frac{EA\Delta_{AB}}{L}\cos\alpha$$

$$N_{AD} = \frac{\partial U}{\partial \Delta_{AD}} = \frac{EA\Delta_{AD}}{L}\cos\alpha$$

From the geometry, the elongation of AC, AB and AD are, respectively, Δ, $\Delta_{AB} = \Delta \cos\alpha$ and $\Delta_{AD} = \Delta \cos\alpha$. Thus,

$$N_{AB} = \frac{EA\Delta_{AB}}{L}\cos\alpha = \frac{EA\Delta}{L}\cos^2\alpha$$

$$N_{AD} = \frac{EA\Delta_{AD}}{L}\cos\alpha = \frac{EA\Delta}{L}\cos^2\alpha$$

Applying the method of joint to A yields

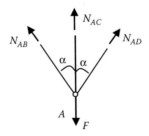

$$N_{AC} + N_{AB}\cos\alpha + N_{AD}\cos\alpha = F$$
$$\frac{EA\Delta}{L} + \frac{EA\Delta}{L}\cos^3\alpha + \frac{EA\Delta}{L}\cos^3\alpha = F$$

$$\Delta = \cfrac{F}{\cfrac{EA}{L} + \cfrac{EA}{L}\cos^3\alpha + \cfrac{EA}{L}\cos^3\alpha}$$

$$= \cfrac{F}{\cfrac{EA}{L} + 2\cfrac{EA}{L}\cos^3\alpha}$$

Substituting the above Δ into the expressions of the three member forces yields

$$N_{AC} = \frac{EA\Delta}{L} = \frac{F}{1 + 2\cos^3\alpha}$$

$$N_{AB} = N_{AD} = \frac{EA\Delta}{L}\cos^2\alpha = \frac{F\cos^2\alpha}{1 + 2\cos^3\alpha}$$

EXAMPLE 11.2

A cantilever beam supports a uniformly distributed load, q. Use Castigliano's second theorem to determine the deflections at points A and B (E and I are constant).

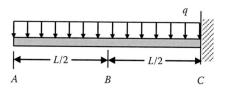

Solution

To use Castigliano's second theorem, imaginary forces F_A and F_B are applied at points A and B, respectively, in the calculation of strain energy. To remove them, these imaginary forces will be replaced by zeros after the derivatives have been taken.

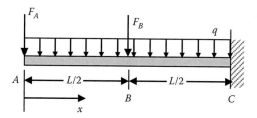

The strain energy for a beam is (Table 11.1)

$$U = \frac{1}{2}\int_0^L \frac{M^2(x)}{EI}\,dx$$

The bending moment due to the applied distributed load and the two imaginary forces is (see Section 6.2.3)

$$M(x) = -F_A\langle x - 0\rangle - F_B\left\langle x - \frac{L}{2}\right\rangle - \frac{1}{2}q\langle x - 0\rangle^2$$

$$= -F_A x - F_B\left\langle x - \frac{L}{2}\right\rangle - \frac{1}{2}qx^2$$

By Castigliano's second theorem (Equation 11.7), the vertical deflections at A and B are, respectively,

$$\delta_A = \frac{\partial U}{\partial F_A} = \int_0^L \frac{M(x)}{EI}\frac{\partial M(x)}{\partial F_A}\,dx$$

Introduce $F_A = F_B = 0$ after the derivative

$$= \int_0^L \frac{-P_A x - P_B\langle x - L/2\rangle - qx^2/2}{EI}(-x)\,dx$$

$$= \int_0^L \frac{qx^2/2}{EI}x\,dx = \frac{qL^4}{8EI}$$

$$\delta_B = \frac{\partial U}{\partial F_B} = \int_0^L \frac{M(x)}{EI}\frac{\partial M(x)}{\partial F_B}\,dx$$

Introduce $F_A = F_B = 0$ after the derivative

$$= \int_0^L \frac{-P_A x - P_B\langle x - L/2\rangle - qx^2/2}{EI}\times(-1)\left\langle x - \frac{L}{2}\right\rangle\,dx$$

$$= \int_0^L \frac{qx^2/2}{EI}\langle x - L/2\rangle\,dx$$

$$= \int_{L/2}^L \frac{qx^2/2}{EI}\langle x - L/2\rangle\,dx = \frac{17qL^4}{384}$$

The above deflections can be verified by the formulas in Table 6.2.

EXAMPLE 11.3

Use the unit load method to find the deflections at A and B of the beam shown in figure below.

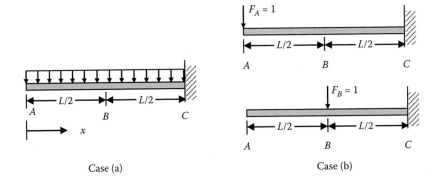

Case (a)   Case (b)

Solution

To apply Equation 11.10 to solve this problem, case (a) is set as the beam subject to the uniformly distributed load and case (b) is set as the beam subject to a unit downward force applied at points A and B, respectively.

The bending moments of cases (a) and (b) can be calculated and introduced into Equation 11.10 to compute the deflections.

The bending moment of case (a) is

$$M_a(x) = -\frac{qx^2}{2}$$

The bending moments of case (b) are, respectively,

$$M_b(x) = -F_A \times x = -x$$

when only F_A is applied, and

$$M_b(x) = 0, \quad 0 \le x \le L/2$$
$$M_b(x) = -F_B \times (x - L/2) = -(x - L/2), \quad L/2 \le x \le L$$

when only F_B is applied. Thus,

$$\delta_A = \int_L \frac{M_a(x)M_b(x)}{EI}\,dx = \int_L \frac{(-qx^2/2)(-x)}{EI}\,dx = \frac{qL^4}{8EI}$$

$$\delta_B = \int_L \frac{M_a(x)M_b(x)}{EI}\,dx = \int_0^{L/2} \frac{(-qx^2/2)(0)}{EI}\,dx + \int_{L/2}^L \frac{(-qx^2/2)(-x + L/2)}{EI}\,dx$$

$$= \frac{17qL^4}{384EI}$$

The results are identical to the solutions from Example 11.2.

EXAMPLE 11.4

The plane frame structure is loaded as shown in figure below. Determine the horizontal displacement, the vertical deflection and the angle of rotation of the section at C. The stiffness of the two members, EI, is constant. Ignore axial and shear deformation of the members.

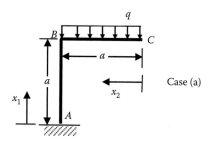

Case (a)

Solution

The horizontal displacement, the vertical deflection and angle of rotation at C can be determined by applying a unit horizontal force, a unit vertical force and a unit bending moment, respectively, at C. Since both the axial and the shear deformation are ignored, only the strain energy due to bending is required when applying the unit load method. In Equation 11.9, the system shown in figures above and below is taken as case (a), and case (b) is taken as follows:

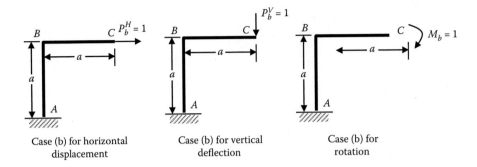

Case (b) for horizontal   Case (b) for vertical   Case (b) for
displacement   deflection   rotation

Since the frame consists of members of different orientations, local coordinates, x_1 for the column and x_2 for the beam, are set to simplify calculation of the bending moment.

	Bending moment in the column from A to B $0 \le x_1 \le a$	Bending moment in the beam from C to B $0 \le x_2 \le a$
Case (a) figure above	$M_a(x) = -\dfrac{qa^2}{2}$	$M_a(x) = -\dfrac{qx_2^2}{2}$
Case (b) horizontal displacement at C	$M_b(x) = -P_b^H(a - x_1) = -(a - x_1) = -(a - x_1)$	0
Case (b) deflection at C	$M_b(x) = -P_b^V a = -a$	$M_b(x) = -P_b^V x_2 = -x_2$
Case (b) rotation at C	$M_b(x) = -M_b = -1$	$M_b(x) = -M_b = -1$

Introducing the above bending moments into Equation 11.9 yields the following:
The horizontal displacement at C is

$$d_C^H = \int_L \frac{M_a(x)M_b(x)}{EI}\,dx = \int_{CB} \frac{(-qx_2^2/2) \times (0)}{EI}\,dx_2 + \int_{AB} \frac{(-qa^2/2) \times (x_1 - a)}{EI}\,dx_1$$

$$= \frac{qa^4}{4EI}$$

The vertical deflection at C is

$$d_C^V = \int_L \frac{M_a(x)M_b(x)}{EI}\,dx = \int_{CB} \frac{(-qx_2^2/2) \times (-x_2)}{EI}\,dx_2 + \int_{AB} \frac{(-qa^2/2) \times (-a)}{EI}\,dx_1$$

$$= \frac{5qa^4}{8EI}$$

The angle of rotation at C is

$$\theta_C = \int_L \frac{M_a(x)M_b(x)}{EI}\,dx = \int_{CB} \frac{(-qx_2^2/2) \times (-1)}{EI}\,dx_2 + \int_{AB} \frac{(-qa^2/2) \times (-1)}{EI}\,dx_1$$

$$= \frac{2qa^3}{3EI}$$

EXAMPLE 11.5

An aluminium wire 7.5 m in length with a cross-sectional area of 100 mm² is stretched between a fixed pin and the free end of the cantilever as shown in figure below. The beam is subjected to a uniformly distributed load of 12 kN/m. Young's modulus and second moment of area of the beam are, respectively, 200 GPa and 20×10^6 mm⁴. Determine the force in the wire.

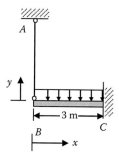

Solution

This is a statically indeterminate structure of first order. The joint at A can be released and replaced by an unknown axial force F_A acting at A. The strain energy of the system can then be calculated in terms of the applied distributed load and the unknown axial force. From the Castigliano's second theorem (Equation 11.7), the derivative of the strain energy with respect to F_A yields the displacement of A in the vertical direction. This vertical displacement must be zero since the point is pinned to the ceiling, which provides the following equation for the solution of the unknown axial force, F_A

$$\frac{\partial U}{\partial F_A} = 0$$

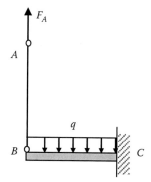

The axial force in the wire is F_A and the strain energy in AB is

$$U_{AB} = \frac{1}{2} \int_A^B \frac{F_A^2}{EA} \, dx = \frac{1}{2} \frac{F_A^2 L_{AB}}{EA}$$

The bending moment in the beam is $F_A x - (1/2)qx^2$ and the strain energy in BC is

$$U_{BC} = \frac{1}{2} \int_B^C \frac{M(x)^2}{EI} \, dx = \frac{1}{2} \int_B^C \frac{(F_A x - qx^2/2)^2}{EI} \, dx$$

Then, $U = U_{AB} + U_{BC}$ and

$$\frac{\partial U}{\partial F_A} = \frac{\partial U_{AB}}{\partial F_A} + \frac{\partial U_{BC}}{\partial F_A} = \frac{F_A L_{AB}}{EA} + \int_B^C \frac{(F_A - qx^2/2)}{EI} x\, dx$$

$$= \frac{F_A L_{AB}}{EA} = \frac{1}{EI}\left[\frac{F_A L_{BC}^2}{2} - \frac{q L_{BC}^2}{8}\right] = 0$$

So

$$\frac{F_A \times 7.5 \text{ m}}{70 \times 10^9 \text{ N/m}^2 \times 100 \times 10^{-6} \text{ m}^2} + \frac{F_A \times (3 \text{ m})^2/2}{200 \times 10^9 \text{ N/m}^2 \times 20 \times 10^{-6} \text{ m}^4}$$

$$= \frac{12 \times 10^3 \text{ N/m} \times (3 \text{ m})^4/8}{200 \times 10^9 \text{ N/m}^2 \times 20 \times 10^{-6} \text{ m}^4}$$

which yields

$$F_A = 9.145 \text{ kN}$$

The axial force in the wire is 9.145 kN.

EXAMPLE 11.6

To determine the deflection at A of the beam (figure below) loaded with a point force P and a bending moment PL, the following solution is obtained by using Castigliano's second theorem. Is the solution correct and why?

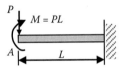

The bending moment of the beam is

$$M(x) = PL - Px = P(L - x)$$

From Equation 10.7

$$\delta_B = \frac{\partial U}{\partial P} = \int_0^L \frac{M(x)}{EI}\frac{\partial M(x)}{\partial P}\, dx = \int_0^L \frac{P(L-x)}{EI}(L-x)\, dx$$

$$= \frac{PL^3}{3EI}$$

Solution

The question tests your understanding of Castigliano's second theorem. The strain energy in Equation 11.7 must be expressed in terms of the applied loads that must be considered as independent forces. In this question, the applied bending moment is related to the applied point force by P. Thus, the derivative with respective to P is taken in relation to not just the force, but also the applied moment. The solution shown above is, therefore, not correct. The correct solution can be obtained by expressing the strain energy in terms of the force P and a bending moment, MP, which is considered completely independent of P. After the derivatives with respect to P is taken, the bending moment is replaced by PL to obtain the deflection.

The bending moment of the beam is

$$M(x) = M_P - Px$$

From Equation 11.7

$$\delta_B = \frac{\partial U}{\partial P} = \int_0^L \frac{M(x)}{EI} \frac{\partial M(x)}{\partial P}\, dx = \int_0^L \frac{M_P - Px}{EI}(-x)\, dx$$

$$= \int_0^L \frac{PL - Px}{EI}(-x)\, dx - \frac{PL^3}{6EI} \quad (M_P \text{ is replaced by } PL \text{ here})$$

Because the final solution is negative, the deflection at A is in the opposite direction of the applied force P.

EXAMPLE 11.7

Find the vertical deflection of point E in the pin-jointed steel truss shown in figure below due to the applied loads at B and F. EA is constant for all members.

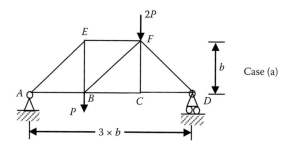

Case (a)

Solution

This is a typical example showing how the unit load method can be applied to find deflection of a truss system. Since this is a pin-jointed structure subjected to loads applied through joints only, within each member, only a constant axial force exists and Equation 11.11 applies. Thus, the system shown in figures above and below is taken as case (a) and the following system is taken as case (b), where an imaginary unit force is applied vertically at E.

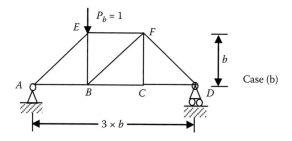

Case (b)

To use Equation 11.11, the axial forces of all the members for cases (a) and (b) must be calculated first. This can be easily done by the method of joint or/and the method of section.

The calculation of the axial forces and the deflection by Equation 11.11 can be presented in the following tabular form.

Member	Length	N_a Case (a)	N_b Case (b)	$N_a \times N_b \times L$
AB	b	$\dfrac{4P}{3}$	$\dfrac{2}{3}$	$\dfrac{8Pb}{9}$
AE	$\sqrt{2}b$	$-\dfrac{4\sqrt{2}P}{3}$	$-\dfrac{2\sqrt{2}}{3}$	$\dfrac{16\sqrt{2}Pb}{9}$
BC	b	$\dfrac{5P}{3}$	$\dfrac{1}{3}$	$\dfrac{5Pb}{9}$
BF	$\sqrt{2}b$	$-\dfrac{\sqrt{2}P}{3}$	$\dfrac{\sqrt{2}}{3}$	$-\dfrac{2\sqrt{2}Pb}{9}$
BE	b	$\dfrac{4P}{3}$	$-\dfrac{1}{3}$	$-\dfrac{4Pb}{9}$
CD	b	$\dfrac{5P}{3}$	$\dfrac{1}{3}$	$\dfrac{5Pb}{9}$
CF	b	0	0	0
DF	$\sqrt{2}b$	$-\dfrac{5\sqrt{2}P}{3}$	$-\dfrac{\sqrt{2}}{3}$	$\dfrac{10\sqrt{2}Pb}{9}$
EF	b	$-\dfrac{4P}{3}$	$-\dfrac{2}{3}$	$\dfrac{8Pb}{9}$
$\displaystyle\sum_{\text{all members}} \dfrac{N_a N_b L}{EA}$				$6.22\dfrac{Pb}{EA}$

The vertical deflection at point E is $6.22Pb/EA$ downwards.

EXAMPLE 11.8

Consider the same truss system shown in figure below. Calculate the increase of the distance between points C and E.

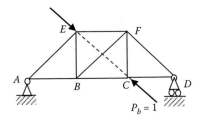

Solution

This question asks for relative displacement of points C and E. Instead of applying unit load at C and E and carrying out respective calculations for case (b), a pair of unit loads is applied simultaneously as shown in figure above. After calculating the axial forces of figure above and following the same procedure of Example 11.7, the relative displacement of C and E, that is, the increase of distance between the two points, is obtained.

Replacing the axial forces in the column of N_b (in Example 11.7) by the respective axial forces calculated from figure above yields the following:

Member	Length	N_a	N_b	$N_a \times N_b \times L$
AB	b	$\dfrac{4P}{3}$	0	0
AE	$\sqrt{2}b$	$-\dfrac{4\sqrt{2}P}{3}$	0	0
BC	b	$\dfrac{5P}{3}$	$\dfrac{\sqrt{2}}{2}$	$\dfrac{5\sqrt{2}Pb}{6}$
BF	$\sqrt{2}b$	$-\dfrac{\sqrt{2}P}{3}$	-1	$\dfrac{2Pb}{3}$
BE	b	$\dfrac{4P}{3}$	$\dfrac{\sqrt{2}}{2}$	$\dfrac{2\sqrt{2}Pb}{3}$
CD	b	$\dfrac{5P}{3}$	0	0
CF	b	0	$\dfrac{\sqrt{2}}{2}$	0
DF	$\sqrt{2}b$	$-\dfrac{5\sqrt{2}P}{3}$	0	0
EF	b	$-\dfrac{4P}{3}$	$\dfrac{\sqrt{2}}{2}$	$-\dfrac{2\sqrt{2}Pb}{3}$
$\displaystyle\sum_{\text{all members}} \dfrac{N_a N_b L}{EA}$				$1.85\dfrac{Pb}{EA}$

The distance between C and E is increased by $1.85Pb/EA$.

11.6 CONCEPTUAL QUESTIONS

1. Define 'strain energy' and derive a formula for it in the case of a uniform bar in tension.
2. Can strain energy be negative?
3. Can virtual strain energy be negative?
4. When a linearly elastic structure is subjected to more than one load, can the strain energy stored in the structure due to the applied loads be calculated by superposition

of the strain energy of the structure under the action of the loads applied individually? Discuss the following two cases (figure below).

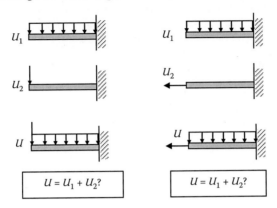

5. Explain how the deflection of beam under a single point load can be found by a strain energy method.
6. State and explain Castigliano's first theorem.
7. State and explain Castigliano's second theorem. How can it be used to determine support reactions of a structure?
8. What are meant by the terms 'virtual force', 'virtual displacement' and 'virtual work'?
9. What is the difference between the work done by a real force and that by a virtual force?
10. Can virtual work be negative?
11. When the unit load method is used to determine the deflection at a point and the calculated deflection is negative, explain why this happens and what the direction of the deflection is.
12. Explain the principle of virtual work and how it can be used in structural and stress analysis.

11.7 MINI TEST

PROBLEM 11.1

The beams shown in figure below are subjected to a combined action of a force and a moment/couple. Can the strain energy under the combined action be calculated by superposition of the strain energy stored in the beam due to the action of the force and the moment applied separately?

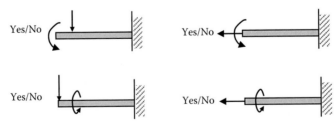

PROBLEM 11.2

The beam shown in figure below is subjected to two identical point forces applied at A and B. The strain energy of the beam, U, is expressed in terms of P, and the derivative of U with respect to P, $\partial U/\partial P$, is

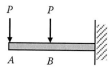

1. The deflection at A.
2. The deflection at B.
3. The average deflection at A and B.
4. The total deflection at A and B.

PROBLEM 11.3

Determine the deflection at A and rotation at B of the beam shown in figure below. EI is a constant.

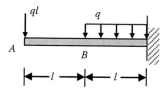

PROBLEM 11.4

Determine the deflection at G of the truss shown in figure below, using the unit load method. The top and bottom members are made of timber with $E_{tb} = 10$ GPa and $A_{tb} = 200$ cm². The diagonal members are also made of timber with $E_d = 10$ GPa and $A_d = 80$ cm². The vertical members are made of steel with $E_{steel} = 20$ GPa. The cross-sectional areas of the vertical members are $A_v = 1.13$ cm² except the central one, whose cross-sectional area is doubled.

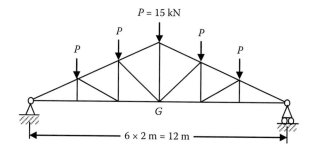

PROBLEM 11.5

Find the vertical deflection at point B in the pin-jointed steel truss shown in figure below due to the applied load at B. Let $E = 200$ GPa. Use Castigliano's second theorem and the unit load method.

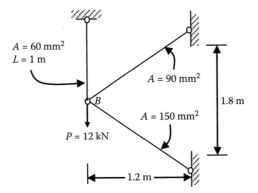

Chapter 12

Bending of thin plates

A plate is a structural component with a thickness very much smaller than other dimensions as shown in Figure 12.1. Normally, a plate is supported along the edges and loads on the plate are applied in the thickness direction. Comparing with bending of beams, a plate will bend in both the x- and y-directions due to the deflection in the z-direction. When a plate is loaded, at an arbitrary point three independent displacements are expected:

- Two in-plane displacements, that is,
  - Displacement in the x-direction, u
  - Displacement in the y-direction, v

- One out-of-plane displacement, that is,
  - Deflection in the z-direction, w

The deflection in the thickness direction, w, is far larger than the two in-plane displacements and, therefore, is the dominating displacement in the analysis of plate bending. How a plate can be better analysed depends on the thickness-to-side ratio, that is, h/a and h/b of Figure 12.1. If a plate is thick (large value of the ratio), a full three-dimensional analysis may be required, while a thin plate (small value of the ratio) requires only a two-dimensional analysis.

12.1 THIN PLATE THEORY

As a rule of thumb, plates with $h/a < 0.2$ and $h/b < 0.2$ fall into thin plate category. When a plate is thin and the deflection of the plate is far smaller than the thickness, the analysis can be simplified considerably. The classic thin plate theory is based on the assumptions given below.

---

a. The plate material is homogeneous, isotropic and elastic.
b. Before loading, the mid-surface of a plate is flat. If this is not the case, significant membrane stresses may be developed in the mid-surface of the plate and the load–deflection relationship may become nonlinear.
c. A plane normal to the mid-surface before bending remains normal to the mid-surface after bending, which is equivalent to the plane section assumption we introduced in Section 5.1 for bending of beams.

---

d. Transverse normal stresses (σ_z) are negligible compared with the two in-plane stresses, σ_x and σ_y, due to the thin thickness.

e. Transverse shear strains (γ_{xz} and γ_{yz}) are very small and, therefore, the associated deformations are neglected. This assumption is directly related to assumption c. If this is violated, the plane section assumption will also be violated.

f. The mid-surface remains a neutral plane during bending, that is, it is not strained ($\varepsilon_x = \varepsilon_y = \gamma_{xy}$).

The above are called Kirchhoff's assumptions that apply for small deflection of thin plates.

Figure 12.2 illustrates the side view of the deflection, the internal forces and stresses of a thin plate subjected to transverse loading. Comparing Figure 12.2 with Figures 5.1 and 5.2, it is apparent that the deformation of a thin plate is similar to the deformation observed in Chapter 5 for bending of beams. For bending of a thin plate, the difference is that we now have to consider bending in both the x- and y-directions, which are interactive due to Poisson's effect.

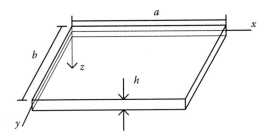

Figure 12.1 A thin rectangular plate.

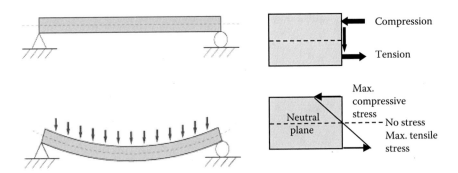

Figure 12.2 Bending of thin plate and bending stresses.

12.2 COMPARISONS OF BENDING OF BEAMS AND BENDING OF THIN PLATES

We are going to describe the governing equations of plate bending in Table 12.1 through comparisons with bending of beams to avoid detailed mathematical derivations of them.

Table 12.1 Bending of plates viz bending of beams

Beam	Plate
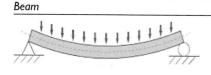 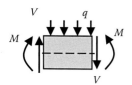	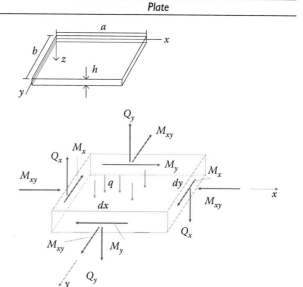
Considering bending relative to the *x*-direction only, i.e., one-way bending	Considering bending relative to both the *x*- and *y*-directions, i.e., two-way bending
(i) Bending moment relative to the *x*-direction $$M_x = -EI\frac{d^2w}{dx^2}$$	(i) Bending moments relative to the *x*- and *y*-directions $$M_x = -D\left(\frac{\partial^2 w}{\partial x^2} + v\frac{\partial^2 w}{\partial y^2}\right)$$ $$M_y = -D\left(\frac{\partial^2 w}{\partial y^2} + v\frac{\partial^2 w}{\partial x^2}\right)$$
The moment is proportional to the curvature of deflection relative to the *x*-direction $$EI = \frac{E}{12}bh^3$$	The moments are proportional to the curvatures of deflection relative to both directions $$D = \frac{Eh^3}{12(1-v^2)}$$
It is calculated over the breath of a beam, *b*	It is called flexural rigidity of plate and is calculated over a unit length (*b* = 1) The second term in the above equations is due to Poisson's effect taking into account the curvature caused by the bending relative to the perpendicular direction Due to that bending occurs simultaneously in both the *x*- and *y*-directions, there exists a coupling effect that is represented a twisting moment, i.e., the twisting moment relative to the *x*–*y* plane $$M_{xy} = -D(1-v)\frac{\partial^2 w}{\partial x\partial y}$$

(Continued)

Table 12.1 (Continued) Bending of plates viz bending of beams

Beam	Plate
(ii) Shear force	(ii) Shear forces
$$V(x) = -EI \frac{d}{dx}\left(\frac{d^2w}{dx^2}\right)$$	$$Q_x = -D \frac{\partial}{\partial x}\left(\frac{\partial^2 w}{\partial x^2} + \frac{\partial^2 w}{\partial y^2}\right)$$ $$Q_y = -D \frac{\partial}{\partial y}\pi\left(\frac{\partial^2 w}{\partial x^2} + \frac{\partial^2 w}{\partial y^2}\right)$$
(iii) Deflection relative to bending in the x-direction	(iii) Deflection relative to bending in both the x- and y-directions
$$EI \frac{d^4w}{dx^4} = q(x)$$	$$D\left(\frac{\partial^4}{\partial x^4} + 2\frac{\partial^4}{\partial x^2 \partial y^2} + \frac{\partial^4}{\partial y^4}\right)w = q(x)$$
	Apparently, bending in both the x- and y-directions and the coupling effect are included

12.3 COMMONLY USED SUPPORT CONDITIONS

In most cases, a plate is supported along its edges. In practice, support can be imposed in various forms, while for design and analysis purpose, they can be mathematically formulated by the following three most commonly used conditions.

12.3.1 Clamped/fixed edges

In this case, an edge of the plate is rigidly fixed to eliminate all displacements and rotations of the edge during bending. To apply this support condition, we nullify the deflections and rotations of the supported edge. Take the fully clamped rectangular plate shown in Figure 12.3 as an example.

At $x = 0$ and $x = a$,

$$w = 0 \quad \text{and} \quad \frac{\partial w}{\partial x} = 0 \text{ (rotation about the x-edge)}$$

At $y = 0$ and $y = b$,

$$w = 0 \quad \text{and} \quad \frac{\partial w}{\partial y} = 0 \text{ (rotation about the y-edge)}$$

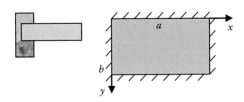

Figure 12.3 Plate with clamped edges.

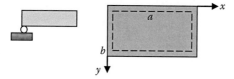

Figure 12.4 Plate with simply supported edges.

12.3.2 Simply supported edges

In this case, an edge of the plate is restrained to prevent any deflection and the edge is allowed to rotate freely during bending. To apply this support condition, we nullify the deflection. The edge must be free of any bending moment to allow free rotation. Take the fully simply supported rectangular plate shown in Figure 12.4 as an example.

At $x = 0$ and $x = a$,

$$w = 0 \quad \text{and} \quad M_x = 0$$

or

$$w = 0 \quad \text{and} \quad M_x = -D\left(\frac{\partial^2 w}{\partial x^2} + v\frac{\partial^2 w}{\partial y^2}\right) = 0$$

At $y = 0$ and $y = b$,

$$w = 0 \quad \text{and} \quad M_y = 0$$

or

$$w = 0 \quad \text{and} \quad M_y = -D\left(\frac{\partial^2 w}{\partial y^2} + v\frac{\partial^2 w}{\partial x^2}\right) = 0$$

12.3.3 Free edges

In this case, an edge of the plate is free to deflect and rotate during bending. Thus, the edge must be free of bending moment and shear force in the thickness direction. Take the rectangular plate shown in Figure 12.5 as an example, where the edges at $x = a$ and $y = b$ are free.

At $x = a$,

$$M_x = -D\left(\frac{\partial^2 w}{\partial x^2} + v\frac{\partial^2 w}{\partial y^2}\right) = 0$$

$$V_x = Q_x + \frac{\partial M_{xy}}{\partial y} \Rightarrow \left[\frac{\partial^3 w}{\partial x^3} + (2-v)\frac{\partial^3 w}{\partial x \partial y^2}\right] = 0$$

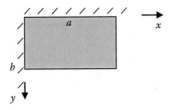

Figure 12.5 Plate with free edges.

At $y = b$,

$$M_y = -D\left(\frac{\partial^2 w}{\partial y^2} + v\frac{\partial^2 w}{\partial x^2}\right) = 0$$

$$V_y = Q_y + \frac{\partial M_{xy}}{\partial x} \Rightarrow \left[\frac{\partial^3 w}{\partial y^3} + (2 - v)\frac{\partial^3 w}{\partial y \partial x^2}\right] = 0$$

The V_x and V_y are the respective equivalent shear forces acting on the planes parallel to the y- and x-axis. They are equal to the respective shear forces, Q_x and Q_y, from Table 12.1 and the shear forces resulting from the twist moments on the planes. As a result, the twist moment along an edge of a thin plate is not an independent boundary condition that must be satisfied in bending analysis of thin plates.

12.4 KEY POINTS REVIEW

- Thin plates are characterised by a flat structural component whose thickness is far smaller than other dimensions. Typically, the largest thickness-to-side ratio is less than 0.2.
- When a plate is considered as a thin plate, any strains relative to the thickness direction can be ignored due to the fact that they are far smaller than other strains.
- The middle plane of a thin plate remains a neutral plane during bending, that is, it is not strained.
- Any plane section perpendicular to the neutral plane remains plane during bending.
- The deflection in the thickness direction is the dominating displacement in the analysis of a thin plate. All other displacement, strains and stresses can be calculated from the deflection.
- The formulation of plate bending is an extension of the formulas for bending of beams in the x-direction to include the bending effect from the y-direction.
- Typical support conditions that are commonly used in bending analysis of plates include simply supported, clamped and free edges.
- Any function that satisfies the governing equation of plate and also the imposed support conditions is the solution of the related plate bending problem.

- Analytical solutions exist only for plates with regular shapes and subjected to simple support conditions. Numerical methods are needed for complex plate problems.
- Apart from the bending moments in the x- and y-directions, M_x and M_y, there exists a twisting moment, M_{xy}, due to the coupling of bending in the two directions.

12.5 EXAMPLES

EXAMPLE 12.1

Find the deflection of the simply supported plate shown in Figure 12.4 when it is subjected to a uniformly distributed pressure, q, from the top surface.

Solution

We need to identify a function for w that can satisfy the simply supported conditions along all the edges, that is, the equations in Section 12.3.1. The function must also satisfy the deflection equation of plate in Table 12.1. Double Fourier series expansion is a good option.

Step 1: Assume

$$w = \sum_{m=1}^{\infty}\sum_{n=1}^{\infty} \omega_{mn} \sin\frac{m\pi x}{a} \sin\frac{n\pi y}{b}$$

where m, n are integers and the ω_{mn} are unknown expansion coefficients.

Step 2: Check against the support conditions by substituting the above w into the moment equations in Table 12.1

$$M_x = -D\left(\frac{\partial^2 w}{\partial x^2} + v\frac{\partial^2 w}{\partial y^2}\right) = D\sum_{m=1}^{\infty}\sum_{n=1}^{\infty}\omega_{mn}\left[\left(\frac{m\pi}{a}\right)^2 + v\left(\frac{n\pi}{b}\right)^2\right]\sin\frac{m\pi x}{a}\sin\frac{n\pi y}{b}$$

$$M_y = -D\left(\frac{\partial^2 w}{\partial y^2} + v\frac{\partial^2 w}{\partial x^2}\right) = D\sum_{m=1}^{\infty}\sum_{n=1}^{\infty}\omega_{mn}\left[\left(\frac{n\pi}{b}\right)^2 + v\left(\frac{m\pi}{a}\right)^2\right]\sin\frac{m\pi x}{a}\sin\frac{n\pi y}{b}$$

Apparently,

$$\sin\frac{m\pi x}{a} = 0 \quad \text{when } x = 0 \text{ or } a$$

$$\sin\frac{n\pi y}{b} = 0 \quad \text{when } y = 0 \text{ or } b$$

then

$$w = M_x = 0 \quad \text{when } x = 0 \text{ or } a$$
$$w = M_y = 0 \quad \text{when } y = 0 \text{ or } b$$

Hence, the simply supported conditions are all satisfied.

Step 3: Check against the deflection equation by substituting the assumed w into the deflection equation in Table 12.1

$$D\left(\frac{\partial^4}{\partial x^4} + 2\frac{\partial^4}{\partial x^2 \partial y^2} + \frac{\partial^4}{\partial y^4}\right)w = q$$

Thus,

$$D\sum_{m=1}^{\infty}\sum_{n=1}^{\infty}\omega_{mn}\left[\left(\frac{m\pi}{a}\right)^4 + 2\left(\frac{m\pi}{a}\right)^2\left(\frac{n\pi}{b}\right)^2 + \left(\frac{n\pi}{b}\right)^4\right]\sin\frac{m\pi x}{a}\sin\frac{n\pi y}{b} = q$$

The constant pressure, q, can be expanded into double Fourier series by following the following mathematical procedure:

$$q(x,y) = \sum_{m=1}^{\infty}\sum_{n=1}^{\infty} q_{mn}\sin\frac{m\pi x}{a}\sin\frac{n\pi y}{b}$$

where

$$q_{mn} = \frac{4}{ab}\int_0^a\int_0^b q(x,y)\sin\frac{m\pi x}{a}\sin\frac{n\pi y}{b}\,dx\,dy$$

After performing the double integration, we have

$$q_{mn} = \begin{cases} \dfrac{16q}{\pi^2 mn} & m,n = 1,3,5,\dots \\ 0 & \text{otherwise} \end{cases}$$

Introducing the expansion of q into the deflection equation yields

$$D\sum_{m=1}^{\infty}\sum_{n=1}^{\infty}\omega_{mn}\left[\left(\frac{m\pi}{a}\right)^4 + 2\left(\frac{m\pi}{a}\right)^2\left(\frac{n\pi}{b}\right)^2 + \left(\frac{n\pi}{b}\right)^4\right]\sin\frac{m\pi x}{a}\sin\frac{n\pi y}{b}$$

$$= \sum_{m=1}^{\infty}\sum_{n=1}^{\infty}\frac{16q}{\pi^2 mn}\sin\frac{m\pi x}{a}\sin\frac{n\pi y}{b}$$

A comparison of both sides of the above equation results in

$$\omega_{mn} = \frac{16q}{D\pi^2 mn[(m\pi/a)^4 + 2(m\pi/a)^2(n\pi/b)^2 + (n\pi/b)^4]} \qquad m,n = 1,3,5,\dots$$

If ω_{mn} takes the above form, the deflection equation of plate in Table 12.1 can be satisfied. Therefore, the final solution of the deflection is

$$w(x,y) = \sum_{m}^{\infty} \sum_{n}^{\infty} \left\{ \frac{16q_0}{D\pi^2 mn[(m\pi/a)^4 + 2(m\pi/a)^2(n\pi/b)^2 + (n\pi/b)^4]} \right\}$$
$$\sin\frac{m\pi x}{a}\sin\frac{n\pi y}{b} \quad m,n = 1,3,5,\ldots$$

Once the deflection is found, the bending moments, twist moment and shear forces can be calculated by using the relative equations in Table 12.1.

EXAMPLE 12.2

Find the deflection of the plate shown in figure below. The plate is simply supported along the two opposite edges at $x = 0$ and $x = a$, and subjected to a distributed pressure $q(x,y)$.

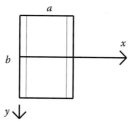

Solution

The plate is simply supported along $x = 0$ and $x = a$. From Example 12.1, we know that using a sinusoidal function in the x-direction can satisfy the simply supported conditions.

Step 1: Assume

$$w = \sum_{m=1}^{\infty} Y_m(y)\sin\frac{m\pi x}{a}$$

where $Y_m(y)$ is an arbitrary function of y.

Step 2: Check against the support conditions at $x = 0$ and $x = a$

$$M_x = -D\left(\frac{\partial^2 w}{\partial x^2} + v\frac{\partial^2 w}{\partial y^2}\right) = -D\sum_{m=1}^{\infty}\left[v\frac{d^2 Y_m}{dy^2} - \left(\frac{m\pi}{a}\right)^2 Y_m(y)\right]\sin\frac{m\pi y}{a}$$

Since

$$\sin\frac{m\pi x}{a} = 0 \quad \text{when } x = 0 \text{ or } a$$

$$w = M_x = 0 \quad \text{when } x = 0 \text{ or } a$$

Step 3: Check against the deflection equation by introducing the assumed w into the equation in Table 12.1

$$D\left(\frac{\partial^4}{\partial x^4} + 2\frac{\partial^4}{\partial x^2 \partial y^2} + \frac{\partial^4}{\partial y^4}\right)w = q$$

$$\sum_{m=1}^{\infty}\left[\frac{d^4Y_m}{dy^4} - 2\left(\frac{m\pi x}{a}\right)^2 \frac{d^2Y_m}{dy^2} + \left(\frac{m\pi x}{a}\right)^4 Y_m\right]\sin\frac{m\pi x}{a} = \frac{q}{D}$$

Step 4: Expand $q(x,y)/D$ into Fourier series

$$\frac{q}{D} = \sum_{m=1}^{\infty} q_m \sin\frac{m\pi x}{a}$$

where

$$q_m = \frac{2}{a}\int_0^a \frac{q}{D}\sin\frac{m\pi x}{a}\,dx$$

Thus,

$$\sum_{m=1}^{\infty}\left[\frac{d^4Y_m}{dy^4} - 2\left(\frac{m\pi x}{a}\right)^2 \frac{d^2Y_m}{dy^2} + \left(\frac{m\pi x}{a}\right)^4 Y_m\right]\sin\frac{m\pi x}{a} = \sum_{m=1}^{\infty} q_m \sin\frac{m\pi x}{a}$$

A comparison of both sides of the equation yields

$$\frac{d^4Y_m}{dy^4} - 2\left(\frac{m\pi}{a}\right)^2 \frac{d^2Y_m}{dy^2} + \left(\frac{m\pi}{a}\right)^4 Y_m = q_m$$

The solution of the above differential equation is

$$Y_m = A_m \cosh\frac{m\pi y}{a} + B_m \frac{m\pi y}{a}\sinh\frac{m\pi y}{a} + C_m \sinh\frac{m\pi y}{a} + D_m \frac{m\pi y}{a}\cosh\frac{m\pi y}{a} + f_m(y)$$

where $f_m(y)$ is a particular solution depending on q_m. A_m, B_m, C_m and D_m are determined from imposing the support conditions at the other two edges. This example shows that if a rectangular plate is simply supported along any two opposite edges, analytical solutions are possible no matter how the other two edges are supported.

EXAMPLE 12.3

Figure below shows part of a thin and long elastic plate, which is simply supported on the edges $y = 0$ and $y = +a$ and is continuous over a series of rigid beams spaced a distance a apart in the x direction. The plate carries the load distribution $q_0 \sin(\pi y/a)$. The flexural rigidity of the plate is denoted by D and Poisson's ratio is 0.3.

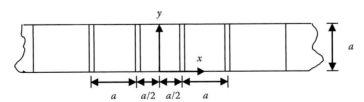

a. State the boundary conditions for a typical panel of the plate.
b. Given that the deflection of a typical panel may be represented by the following expression, determine the constants A and B:

$$w = \frac{q_0 a^4}{D\pi^4}\left(1 + A\cosh\frac{\pi x}{a} + Bx\sinh\frac{\pi x}{a}\right)\sin\frac{\pi y}{a}$$

c. Calculate the deflection at the point $x = 0$, $y = a/2$.
d. Calculate the value of M_x at point $x = 0$, $y = a/2$ and $x = a/2$, $y = a/2$. Sketch the distribution of M_x along the line $y = a/2$.

Solution

The panel continuous over a number of spans with equally spaced rigid beam supports. The load is constant in the x-direction and varies sinusoidally in the y-direction. This implies that each of the panels can be approximately taken as an independent plate with proper supports introduced at the connections along the rigid beam.

Step 1 (answer to question a): Define the support conditions
  Along the long edge, the panel is simply supported, that is,

At $y = 0$ and $y = a : w = 0$ and $M_y = 0$

Since the plate rests on rigid beams, the deflection of the plate along these beams is zero. The plate is long and the loading is uniform in the x-direction, resulting in zero rotation of the plate sections along the support beams. Hence, the beams provide equivalent fixed supports to each of the individual panels. For the central panel shown in Figure 12.3, for example, we have

At $x = \pm a/2 : w = 0$ and $\dfrac{\partial w}{\partial x} = 0$

  We now analyse a plate with two opposite edges simply supported and the other two edges clamped.

Step 2 (answer to question b): For the central panel, check whether the given form of deflection satisfies the support conditions defined in Step 1. Apparently, from the given deflection

at $y = 0$ and $y = a$

  $w = 0$

  $M_y = -D\left[\dfrac{\partial^2 w}{\partial y^2} + \nu\dfrac{\partial^2 w}{\partial x^2}\right] = 0$

are automatically satisfied.

  At $x = \pm a/2$

$w = 0$, which requires

$$1 + A\cosh\frac{\pi}{2} + B\frac{a}{2}\sinh\frac{\pi}{2} = 0$$

or

$$1 + 2.509A + 1.15aB = 0$$

$\partial w/\partial x = 0$, which requires

$$A\frac{\pi}{a}\cosh\frac{\pi}{2} + B\left(\sinh\frac{\pi}{2} + \frac{\pi}{2}\cosh\frac{\pi}{2}\right) = 0$$

or

$$\frac{7.23}{a}A + 6.24B = 0$$

Solution of the above equations yields

$$A = -0.85 \quad \text{and} \quad Ba = 0.99$$

Step 3 (answer to question c): Deflection at $x = 0$ and $y = a/2$
The deflection of the panel is

$$w = \frac{q_0 a^4}{D\pi^4}\left(1 - 0.85\cosh\frac{\pi x}{a} + \frac{0.99}{a}x\sinh\frac{\pi x}{a}\right)\sin\frac{\pi y}{a}$$

at $x = 0$ and $y = a/2$

$$w = \frac{q_0 a^4}{D\pi^4}[1 - 0.85] = 0.15\frac{q_0 a^4}{D\pi^4}$$

Step 4 (answer to question d): Bending moment at $x = 0, y = a/2$ and $x = a/2, y = a/2$

$$\frac{\partial^2 w}{\partial x^2} = \frac{q_0 a^4}{D\pi^4}\sin\frac{\pi y}{a}\left[\left(A\left(\frac{\pi}{a}\right)^2 + 2B\left(\frac{\pi}{a}\right)\right)\cosh\frac{\pi x}{a} + B\left(\frac{\pi}{a}\right)^2 x\sinh\frac{\pi x}{a}\right]$$

$$\frac{\partial^2 w}{\partial y^2} = -\frac{q_0 a^4}{D\pi^4}\sin\frac{\pi y}{a}\left[1 + A\cosh\frac{\pi x}{a} + Bx\sinh\frac{\pi x}{a}\right]\left(\frac{\pi}{a}\right)^2$$

$$M_x = -D\left(\frac{\partial^2 W}{\partial x^2} + v\frac{\partial^2 W}{\partial y^2}\right)$$

- At $x = 0$ and $y = a/2$

$$\frac{\partial^2 w}{\partial x^2} = \frac{q_0 a^4}{D\pi^4}\left(A\left(\frac{\pi}{a}\right)^2 + 2B\left(\frac{\pi}{a}\right)\right) = -0.22\frac{q_0 a^2}{D\pi^2}$$

$$\frac{\partial^2 w}{\partial y^2} = -\frac{q_0 a^4}{D\pi^4}\left(\frac{\pi}{a}\right)^2(1 + A) = -0.15\frac{q_0 a^2}{D\pi^2}$$

$$M_x = +0.265\frac{q_0 a^2}{D\pi^2}$$

- At $x = a/2$ and $y = a/2$

$$\frac{\partial^2 w}{\partial x^2} = \frac{q_0 a^4}{D\pi^4}\left\{2.509\left[A\left(\frac{\pi}{a}\right)^2 + 2B\left(\frac{\pi}{a}\right)\right] + 2.301B\left(\frac{\pi}{a}\right)^2\frac{a}{2}\right\} = 0.59\frac{q_0 a^2}{D\pi^2}$$

$$\frac{\partial^2 w}{\partial y^2} = -\frac{q_0 a^4}{D\pi^4}\left[1 + 2.509A + \frac{0.99 \times 2.301}{2}\left(\frac{\pi}{a}\right)^2\right] = 0$$

$$M_x = -0.59\frac{q_0 a^2}{D\pi^2}$$

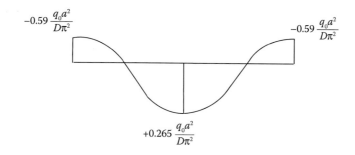

$-0.59\dfrac{q_0 a^2}{D\pi^2}$ $-0.59\dfrac{q_0 a^2}{D\pi^2}$

$+0.265\dfrac{q_0 a^2}{D\pi^2}$

12.6 CONCEPTUAL QUESTIONS

1. What is meant by 'plate' used in structural engineering?
2. What is meant by 'thin plate'?
3. What are the basic assumptions introduced to simplify the solution of plate bending problems?
4. What are the simply supported and clamped boundary conditions for thin plates?
5. How is the twisting moment considered when a thin plate has free edges?
6. Explain the differences and similarities between bending of beams and plates.
7. If a rectangular plate bends relative to only one direction, for example, the x-direction, what observation can you get from a study of the formulas presented in Table 12.1?
8. Is it possible to obtain analytical solutions for simply supported plates subjected to concentrated point loads?
9. Can the double Fourier series method be used directly for clamped or free edge support conditions and why?
10. Can you sketch the distribution of the bending stresses on a section of a thin plate?

12.7 MINI TEST

PROBLEM 12.1

Consider a floor slab occupying the region. $0 \le x \le a$, $0 \le y \le b$, $-h/2 \le z \le h/2$. The slab is simply supported on all edges, and loaded with sand in such a way that the load can be approximated by

$$p_z(x,y) = p_0 \sin\frac{\pi x}{a}\sin\frac{\pi y}{b}$$

Determine the location and magnitude of the maximum deflection, the maximum bending moments and the maximum shear forces in each direction.

PROBLEM 12.2

Describe the assumptions of the thin plate theory. The deflection function for laterally loaded square slab of side a is given by

$$W(x,y) = W_0 \cos\frac{\pi x}{a} \cos\frac{\pi y}{a}$$

where x and y are rectangular coordinates measured from a central origin and W_0 is the deflection at the centre. If the flexural rigidity is D and Poisson's ratio is v determine

    a. The load pattern causing the deflected form
    b. The nature of the boundary conditions
    c. The bending moments at the centre of the slab

PROBLEM 12.3

The thin plate shown in figure below is subjected to a lateral point load at B. The plate is simply supported along sides OA and OC, and free of any supports along sides AB and BC. Show that $w = mxy$ is the solution of the problem and calculate the deflection, bending moments and shear forces of the plate.

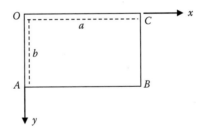

PROBLEM 12.4

The boundary of the elliptical plate shown in figure below is described by the following equation:

$$\frac{x^2}{a^2} + \frac{y^2}{b^2} = 1$$

Show that

$$w = m\left(\frac{x^2}{a^2} + \frac{y^2}{b^2} - 1\right)$$

can be used as the deflection function of the problem and determine the coefficient m. The plate is fully clamped along the boundary, subjected to a lateral pressure q and has a flexural rigidity of D.

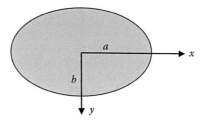

PROBLEM 12.5

Show that the three in-plane stresses can be calculated below

$$\sigma_x = \frac{12M_x z}{h^2}, \quad \sigma_y = \frac{12M_y z}{h^2}, \quad \tau_{xy} = \frac{12M_{xy} z}{h^2}$$

where h is the thickness of plate and z is measured from the mid-surface of the plate.

Chapter 13

Impact loads and vibration

We have studied stresses and strains in structural members caused by statically applied loads. In engineering applications, we may have to deal with structures and loadings of dynamic nature, for example, a moving body or a freely falling mass. On the basis of the principal of conservation of energy, it may be assumed that at the moment a moving body is stopped by a resisting structure causing an impact, its kinetic and potential energies are completely transformed into the internal strain energy of the resisting structure. Hence, there are two major dynamic problems to be solved, that is, for the maximum deformation and stresses of the resisting structure at the moment of impact and the vibration initiated by the impact afterward.

13.1 IMPACT LOAD

The term impact refers to a dynamic effect of a load, which is applied suddenly. For example, a free failing mass strikes the floor of a building. Obviously, the stresses on the floor caused by such a strike are more intensive than the stresses caused if the mass is placed on the floor slowly or statically. Problems involving similar forces may be analysed simply by introducing the following assumptions.

a. Materials are linearly elastic.
b. No energy dissipation takes place during impact.
c. The mass of the structure resisting the impact may be neglected.
d. The deformation of the falling mass is neglected.

On the basis of the principal of energy conservation, at the moment of impact, the total energy of the falling mass is completely transformed into the internal strain energy of the resisting structure. Thus,

$$T + V = U \tag{13.1}$$

where T and V are the respective kinetic and potential energy of the falling mass; U is the increased strain energy of the resisting structure due to the impact.

Table 13.1 Impact factor for some of the most common forms of impact

Impact	Impact factor K_d	
Struck by freely falling mass	$1 + \sqrt{1 + \dfrac{2h}{\Delta_{st}}}$	Falling from a height h
Subjected to suddenly force	2	Equivalent to $h = 0$
Struck vertically by mass with constant velocity	$1 + \sqrt{1 + \dfrac{v^2}{g\Delta_{st}}}$	v is the velocity; g is the gravity
Struck horizontally by mass	$\sqrt{\dfrac{v^2}{g\Delta_{st}}}$	

In most cases, the stress and deformation (σ_{dyna} and Δ_{dyna}) due to impact can be calculated by introducing an impact factor (K_d) that magnifies the stress and deformation (σ_{st} and Δ_{st}) caused by statically applying the mass to the structure. Thus,

$$\sigma_{dyna} = K_d \sigma_{st}$$
$$\Delta_{dyna} = K_d \Delta_{st} \tag{13.2}$$

The values of K_d can be found in Table 13.1.

13.2 VIBRATION

One of the most important dynamic analyses of structures is vibratory motion, or vibration, in which the structure oscillates about a certain static equilibrium position after being set off with an initial and/or subjected to continuous time-dependent input. Typical examples include motion of long span bridges subjected to wind and traffic, earthquakes, offshore structures subjected to waves and factory buildings subjected to machine vibration. Figure 13.1 shows a simple illustration of a single degree-of-freedom (DOF) spring–mass–damper system that oscillates back and forth about its equilibrium position, where k and c are the stiffness of the spring and the damping coefficient of the damper, respectively.

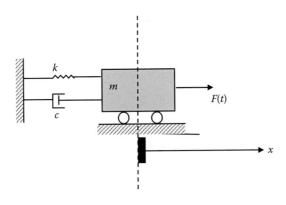

Figure 13.1 Single-degree-of-freedom (SDOF) model.

Important parameters that are used to describe a vibration include the following:

Circular frequency ω_n – is the number of occurrences of a repeating motion per unit time measured by angular displacement covered per unit time. Typical unit of circular frequency is rad/s.

Nature frequency f – is the number of occurrences of a repeating motion per unit time measured by a completed cycle of the repeated motion covered per unit time. Typical unit of natural frequency is Hz (cycles/s).

Period T – is the duration of one complete cycle of motion, and is the reciprocal of the natural frequency, that is, $T = 1/f$. Typical unit of the period is second (s).

Amplitude – is the maximum displacement of a vibratory mass from its static equilibrium position. The unit of amplitude is the same as the unit for displacement.

Vibrations can be broadly classified as *free vibration* or *forced vibration*.

13.2.1 Types of vibration

13.2.1.1 Dynamic equilibrium equation of vibration

Figure 13.2 establishes the dynamic equilibrium of the so-called single-degree-of-freedom (SDOF) system, of which the mass can only move in the horizontal direction.

Figure 13.2 shows the following:

- Moving mass m
- Displacement from static equilibrium position $x(t)$
- Restoring force of spring $F_S(t) = kx(t)$
- Damping force $F_D(t) = c\dot{x}(t)$
- Inertial force $F_I(t) = m\ddot{x}(t)$
- External force input $F(t)$

From the free body diagram of Figure 13.2, we have

$$m\ddot{x} + c\dot{x} + kx = F(t) \tag{13.3}$$

The natural circular frequency or angular velocity of the system is

$$\omega_n = \sqrt{k/m} \tag{13.4}$$

The natural circular frequency can be related to the period (T) and natural frequency (f) of vibration, respectively, by

$$T = \frac{2\pi}{\omega_n} \tag{13.5}$$

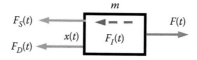

Figure 13.2 Dynamic equilibrium of SDOF mass.

and

$$f = \frac{\omega_n}{2\pi} \tag{13.6}$$

The damping coefficient is found

$$c = 2m\omega_n\xi \tag{13.7}$$

in which

$$\xi = \frac{c}{2m\omega_n} = \frac{c}{2\sqrt{mk}} \tag{13.8}$$

is called ratio of damping or damping ratio of the system.

They follow four different vibration analyses depending on the form of Equation 13.3:

- $F(t) = 0$: Free vibration analysis, that is, once the mass starts moving, there will be no external dynamic forces acting on the system.
- $F(t) \neq 0$: Forced vibration analysis, that is, the mass is subjected to external dynamic forces during the course of vibration.
- $c = 0$: Undamped vibration analysis, that is, the damping effect of the system is small and may be neglected.
- $c \neq 0$: Damped vibration analysis, that is, the damping effect of the system is significant and has to be considered.

13.2.1.2 Free vibration ($F(t) = 0$)

Mathematically, the solution of Equation 13.3 exhibits oscillation of the mass only when the damping ratio, ξ, is smaller than unity.

Figure 13.3 shows the displacement of the mass relative to the equilibrium position when the damping ratio is equal to or greater than unity. The associated systems are, respectively, called critically and over-damped systems. Clearly, there are no oscillations for both cases,

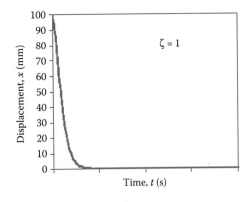
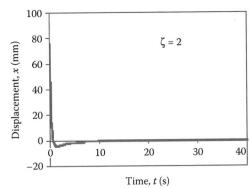

Figure 13.3 Critically and overdamped motion.

and the displacements gradually approach zero. For vibration analysis, we will concentrate on the case when the system is underdamped, that is, when $\xi < 1$.

For an underdamped system under the free vibration condition, the solution of Equation 13.3 is

$$x(t) = Ae^{-\xi\omega_n t}\sin(\omega_d t + \varphi) \tag{13.9}$$

which represents an oscillatory motion of the mass with constant circular frequency ω_d and phase angle φ, but with exponentially decaying amplitude $Ae^{-\xi\omega_n t}$, where $\omega_d = \omega_n\sqrt{1-\xi^2}$, A and φ depend on the initial conditions.

13.2.1.3 Forced vibration (F(t) ≠ 0)

We shall concentrate on a mass–spring–damper system subjected to harmonic excitation. For this simple case, Equation 13.3 may take the following form:

$$m\ddot{x} + c\dot{x} + kx = F_0\sin\omega t \tag{13.10}$$

where F_0 and ω are the amplitude and natural circular frequency of the applied force. The particular or steady-state solution of the above equation is

$$x = \rho\sin(\omega t - \varphi) \tag{13.11}$$

where

$$\rho = \frac{F_0}{k}\frac{1}{\sqrt{[1-(\omega/\omega_n)^2]^2 + (2\xi\omega/\omega_n)^2}} \tag{13.12}$$

$$\varphi = \tan^{-1}\left[\frac{2\xi\omega/\omega_n}{1-(\omega/\omega_n)^2}\right] \tag{13.13}$$

13.2.1.4 Dynamic amplification factor

In Equation 13.12, ρ is the amplitude of the forced vibration and F_0/k is the static displacement (amplitude) of the mass if the maximum force (amplitude of the harmonic excitation) is applied statically to the system. The ratio of the two amplitudes is

$$D = \frac{\rho}{F_0/k} = 1/\sqrt{[1-(\omega/\omega_n)^2]^2 + (2\xi\omega/\omega_n)^2} \tag{13.14}$$

and is called the *dynamic amplification factor* (DAF). The DAF can be used to describe the number of times the displacements of the system caused by the static loads should be multiplied to obtain the displacement when a dynamic load is applied onto the same system.

13.2.1.5 Resonance

Figure 13.4 shows plots of DAF versus ω/ω_n for various values of ξ, which demonstrates that damping tends to diminish amplitude and the peaks of the curves occur at around $\omega/\omega_n = 1$.

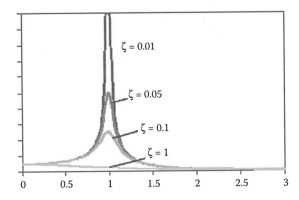

Figure 13.4 DAF viz frequency ratio.

For the undamped vibration, that is, when $\xi = 0$, the DAF approaches infinity, as shown by Equation 13.14. In general, when the frequency ratio approaches unity, the vibration is said to approach a resonance condition characterised by violent oscillation.

13.3 KEY POINTS REVIEW

- At the moment of an impact, the kinetic and potential energies of the moving body are assumed to be transformed to the strain energy stored in the resisting structure.
- The deformation and stresses in the resisting structure due to an impact can be computed by multiplying the deformation and stresses caused by known static loads with an impact factor.
- A system that possesses mass and elasticity can vibrate.
- Initial and/or time-dependent inputs are required to produce vibration.
- Frequency is the number of occurrences of a repeating motion per unit time.
- Frequency can be measured by the angular or linear displacement per unit time.
- Amplitude is a measurement of the maximum displacement of a vibratory mass from its static equilibrium position.
- Damping in the system causes the amplitude of vibration to decrease with time.
- The more damping in the system, the more rapidly the amplitude decay away.
- If there is excessive damping, no vibration will occur.
- At very low driving frequencies (small ω or ω/ω_n), the amplitude of the response is almost the same as the static displacement, F_0/k.
- At very high driving frequencies (large ω or ω/ω_n), the amplitude tends to become zero.
- For underdamped system, the peak amplitude occurs at a frequency ratio very close to unity.
- A driving frequency near to the natural frequency produces a rather critical condition that is called resonance, characterised by violent oscillation of the system.
- The damping in the system can be determined by measuring the rate at which the amplitude decays (see Example 13.4).

13.4 SUMMARY OF THE SOLUTIONS

See Table 13.2.

Table 13.2 Solution of SDOF system

Free vibration (F(t) = 0)	Forced vibration (F(t) ≠ 0)
Undamped c = 0	*Undamped c = 0*
i. Equation	i. Equation
$m\ddot{x} + kx = 0$	$m\ddot{x} + kx = F_0 \sin\omega t$
ii. Solution[a]	ii. Solution[c]
$x = A\sin(\omega_n t + \varphi)$	when $x(0) = \dot{x}(0) = 0$
	$x(t) = \dfrac{F_0}{k}\left[\dfrac{1}{1-(\omega/\omega_n)^2}\right](\sin\omega t - \beta\sin\omega_n t)$
Damped c ≠ 0 and (ξ < 1)	*Damped c ≠ 0 and (ξ < 1)*
i. Equation	i. Equation
$m\ddot{x} + c\dot{x} + kx = 0$	$m\ddot{x} + c\dot{x} + kx = F_o \sin\omega t$
ii. Solution[b]	ii. Particular solution (ξ < 1)[c]
$x = Ae^{-\xi\omega_n t}\sin(\omega_d t + \varphi_d)$	$x = \rho\sin(\omega t - \varphi)$
	$\rho = \dfrac{F_0}{k}\dfrac{1}{\sqrt{[1-(\omega/\omega_n)^2]^2 + (2\xi\omega/\omega_n)^2}}$
	$\varphi = \tan^{-1}\left[\dfrac{2\xi\omega/\omega_n}{1-(\omega/\omega_n)^2}\right]$

[a] The arbitrary constants can be determined by introducing the initial conditions, that is, the known displacements and velocities at $t = 0$.
[b] If the initial conditions are different, the solution will take a different form.
[c] Only the steady-state solution of the system is shown.

13.5 EXAMPLES

EXAMPLE 13.1

Find the maximum deflection and bending stress for the 50 mm × 50 mm steel beam shown in figure below when struck at the mid-span by a 20 kg mass falling from 100 mm above the top surface of the beam. The elastic modulus of the steel is $E = 200$ GPa.

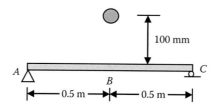

Solution

The deflection and stress of the beam are caused by the falling mass, that is, by the impact load. Therefore, they can be computed by multiplying the static deflection and stress due to applying the mass slowly on the top surface of the beam with a proper impact factor from Table 13.1.

Step 1: Convert the mass into a static load P_{st}

$$P_{st} = 20g = 196\,\text{N}$$

Step 2: Compute the deflection and stress if the mass is statically applied at the mid-span. For a simply supported beam (Table 6.2), the deflection at the mid-span is

$$\Delta_{st} = \frac{P_{st}L^3}{48EI} = \frac{196 \times 1^3}{48 \times 200 \times 10^9 \times (50 \times 10^{-3}) \times (50 \times 10^{-3})^3/12} = 39.2 \times 10^{-6}\,\text{m}$$

The maximum bending stress of the beam due to P_{st} is

$$(\sigma_{st})_{max} = \frac{M_{st}}{I/y_{max}} = \frac{196/4}{[(50 \times 10^{-3}) \times (50 \times 10^{-3})^3/12]/(25 \times 10^{-3})} = 2.4\,\text{MPa}$$

Step 3: Choose an impact factor from Table 13.1. For a falling mass

$$K_d = 1 + \sqrt{1 + \frac{2h}{\Delta_{st}}} = 1 + \sqrt{1 + \frac{2 \times 100 \times 10^{-3}}{39.2 \times 10^{-6}}} = 72.4$$

Step 4: Computer the deflection and stress due to the impact

$$\Delta_d = K_d\Delta_{st} = 72.4 \times 39.2 \times 10^{-6} = 2.84 \times 10^{-3}\,\text{m}$$

$$(\sigma_d)_{max} = K_d \times (\sigma_{st})_{max} = 72.4 \times 2.4 = 173.8\,\text{MPa}$$

It is apparent that significantly larger deflection and stress are caused by the impact load.

EXAMPLE 13.2

A 1 kN motor shown is supported by a cantilever of length L, as shown in figure below. The motor is rotating at a speed of 900 rev/min. The eccentricity of the rotating part generates a centrifugal force that acts periodically at the free end. The steel beam has a section property of $EI = 55 \times 10^4\,\text{N m}^2$. Determine the length of the cantilever at which resonance occurs.

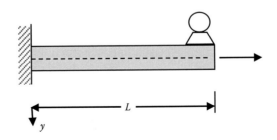

Solution

Resonance occurs when the natural frequency of a vibration system coincides with the frequency of any applied dynamic force. We need, therefore, to know the stiffness, k, of supporting beam so that the natural circular frequency can be calculated. We need also to compute the frequency of the centrifugal force due to the eccentricity for comparison.

Step 1: Calculate the frequency of the vertical component of the centrifugal force

$$\omega = \frac{2\pi \times 900}{60} = 94.25 \, \text{rad/s}$$

Step 2: Calculate the stiffness of the cantilever

For a cantilever, the deflection at the free end due to a static force, P, applied at the same location is (Table 6.2)

$$y_{max} = \frac{PL^3}{3EI}$$

Thus,

$$k = \frac{P}{y_{max}} = \frac{165 \times 10^4}{L^3} \, \text{N/m}$$

The nature circular frequency of the system is

$$\omega_n = \sqrt{\frac{k}{m}} = \sqrt{\frac{165 \times 10^4}{(1000/g)L^3}}$$

When resonance occurs
$\omega_n = \omega$, that is,

$$\sqrt{\frac{165 \times 10^4}{(1000/g)L^3}} = 94.25$$

Then $L = 1.22$ m
Resonance will take place when the length of the cantilever is close to 1.22 m.

EXAMPLE 13.3

Derive the dynamic equilibrium equation of the one-store building shown in figure below. The building is subjected to ground motion $x_g(t)$ during an earthquake. The total stiffness of the supporting columns against horizontal motion of the floor is k. The floor has a mass of m and the damping coefficient of the system is c.

Solution

There are two motions that have to be considered. The total movement of the floor, $x^t(t)$, is the sum of the ground motion, $x_g(t)$, and the relative movement, $x(t)$, of the floor to the ground. Thus,

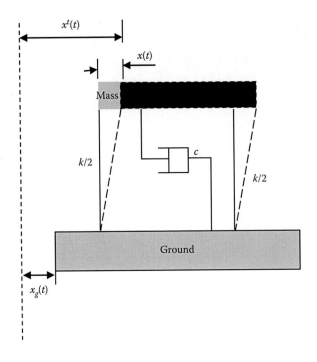

$$x^t(t) = x(t) + x_g(t)$$

The dynamic equilibrium equation of the floor is

$$m\ddot{x}^t(t) + c\dot{x}(t) + kx(t) = 0$$

or

$$m(\ddot{x} + \ddot{x}_g) + c\dot{x} + kx = 0$$

Hence, $m\ddot{x} + c\dot{x} + kx = -m\ddot{x}_g$.

From the right-hand-side term of the above equation, it can be seen that the ground motion generates an inertial force that is the driving force causing vibration of the floor.

EXAMPLE 13.4

At a damped free vibration test, it is found that the amplitude of the system is reduced to the half of its original after five complete cycles of vibration. Determine the ratio of damping of the system.

Solution

Equation 13.9 defines the relationship between the exponentially decaying amplitude and the damping ratio, which can be shown graphically by the figure below.

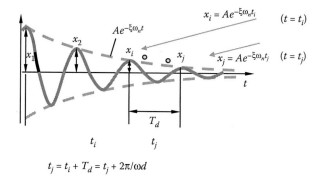

$$t_j = t_i + T_d = t_j + 2\pi/\omega d$$

By comparing the amplitudes at the two consecutive peaks

$$\frac{x_i}{x_j} = \frac{Ae^{-\xi\omega_n t_i}}{Ae^{-\xi\omega_n t_j}} = e^{-\xi\omega_n(t_i - t_j)} = e^{2\pi\xi\omega_n/\omega_d} = e^{2\pi\xi/\sqrt{1-\xi^2}}$$

Taking the natural logarithm of the above ratio,

$$\delta = \ln\left(\frac{x_i}{x_j}\right) = \frac{2\pi\xi}{\sqrt{1-\xi^2}}$$

When the damping ratio is small, $\xi^2 \ll 1$. Thus,

$$\delta = \ln\left(\frac{x_i}{x_j}\right) \approx 2\pi\xi \quad \text{or} \quad \xi \approx \frac{\delta}{2\pi}$$

From the question, we have

$$\frac{x_i}{x_{i+5}} = 2$$

or equivalently

$$\frac{x_i}{x_{i+1}} \times \frac{x_{i+1}}{x_{i+2}} \times \frac{x_{i+2}}{x_{i+3}} \times \frac{x_{i+3}}{x_{i+4}} \times \frac{x_{i+4}}{x_{i+5}} = 2$$

or

$$\ln\left(\frac{x_i}{x_{i+1}} \times \frac{x_{i+1}}{x_{i+2}} \times \frac{x_{i+2}}{x_{i+3}} \times \frac{x_{i+3}}{x_{i+4}} \times \frac{x_{i+4}}{x_{i+5}}\right) = \ln 2$$

or

$$\ln\left(\frac{x_i}{x_{i+1}}\right) + \ln\left(\frac{x_{i+1}}{x_{i+2}}\right) + \ln\left(\frac{x_{i+2}}{x_{i+3}}\right) + \ln\left(\frac{x_{i+3}}{x_{i+4}}\right) + \ln\left(\frac{x_{i+4}}{x_{i+5}}\right) = \ln 2$$

Each of the five terms on the left-hand side of the above equation is the natural logarithm of the ratio of two consecutive peaks and, therefore, is identical to δ. Thus,

$$\delta + \delta + \delta + \delta + \delta = 5\delta = \ln 2$$

From which

$$\delta = \frac{1}{5}\ln 2$$

or

$$\xi = \frac{\delta}{2\pi} = \frac{1}{10\pi}\ln 2 = 0.022$$

Thus, the damping ratio of the system is 0.022.

EXAMPLE 13.5

Consider the beam–machine system shown in figure below. The system consists of a massless beam supporting a machine mass of $m = 2000$ kg, containing an eccentric mass (m_1) of 10 kg with an eccentricity of 0.1 m, which is rotating at a speed of 100 rev/min. Calculate the dynamic amplification that the beam has experienced. The second moment of area of the beam is 4×10^{-6} m^4 and the elastic modulus of the material is $E = 2 \times 10^{11}$ N/m^2. The damping ratio of the system is 0.05.

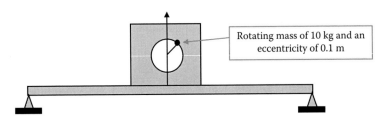

Rotating mass of 10 kg and an eccentricity of 0.1 m

Solution

The vibration of the support beam is due to the rotation of the eccentric mass imposing a period of dynamic force acting vertically on the system. The stiffness of the beam provides spring effect in the vertical direction as in a simple spring–mass system. To solve this problem, we need to first determine the stiffness of the simply supported beam in resisting a force applied at the mid-span. We need also to find the periodical force applied at the same location.

Step 1: Calculate the stiffness of the beam by studying the mid-span deflection of the beam subjected to a concentrated force at the same location. From Table 6.2

$$y_{max} = \frac{PL^3}{48EI}$$

Thus,

$$k = \frac{y_{max}}{P} = \frac{48EI}{L^3} = 3.1 \times 10^5 \, \text{N/m}$$

And the natural circular frequency of the system is

$$\omega_n = \sqrt{\frac{k}{m}} = \sqrt{\frac{3.1 \times 10^5}{2000}} = 12.4 \, \text{rad/s}$$

Step 2: Calculate the circular velocity of the rotating mass, that is, the natural circular frequency of the equivalent periodic force

$$\omega = \frac{2\pi Revs}{60} = \frac{2\pi \times 100}{60} = 10.5 \, \text{rad/s}$$

Step 3: Determine the motion of the rotating mass in the vertical direction. The instantaneous vertical position of the rotating mass, x_1, is the position of the centre of the machine, x, offset by an additional vertical displacement due to the rotation of the mass about the centre. Thus,

$$x_1 = x + 0.1 \sin \omega t$$

Step 4: Write the dynamic equation of the system. The equilibrium is maintained by considering the inertia force of the machine and the rotating mass, the damping effect and the stiffness of the system. It yields

$$(m - m_1)\ddot{x} + m_1\ddot{x}_1 + c\dot{x} + kx = 0$$

or

$$m\ddot{x} + c\dot{x} + kx = 0.1m_1\omega^2 \sin(\omega t)$$

Here $0.1m_1\omega^2$ is the amplitude of the equivalent force due to the rotation of eccentric mass. If the force is applied on the system statically, the beam will deflect by $0.1m_1\omega^2/k$

$$\frac{0.1m_1\omega^2}{k} = \frac{0.1 \times 10 \times 10.5^2}{3.1} = 35.6 \times 10^{-5} \, \text{m}$$

The dynamic amplification is characterised by the DAF of Equation 13.14.

$$D = 1/\sqrt{[1 - (\omega/\omega_n)^2]^2 + (2\xi\omega/\omega_n)^2} = 3.35$$

Thus, the maximum dynamic deflection of the beam is

$$x_{max} = D \times \frac{0.1m_1\omega^2}{k} = 3.35 \times 35.6 \times 10^{-5} = 1.19 \times 10^{-3}\,\text{m}$$

EXAMPLE 13.6

The mass–spring–damper system shown in figure below has the following properties: $m = 2$ kg, $k = 1000$ N/m and $F = 100 \sin (20t)$ N. Determine the damping coefficient such that maximum amplitude of vibration should be less than 0.1 m.

Solution

This is a vibration design problem. A damper can be properly designed to reduce the amplitude of the mass. Apparently, Equation 13.12 can be used to calculate the vibration amplitude for a given value of damping ratio.

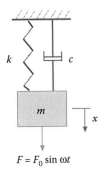

$$F = F_0 \sin \omega t$$

For the system shown in figure above, $F_0 = 100$ N and $\omega = 20$ rad/s

$$\omega_n = \sqrt{\frac{k}{m}} = \sqrt{\frac{1000}{2}} = 22.4\,\text{rad/s}$$

$$\rho = \frac{F_0}{k}\frac{1}{\sqrt{[1 - (\omega/\omega_n)^2]^2 + (2\xi\omega/\omega_n)^2}}$$

$$= \frac{0.1}{\sqrt{0.04 + 3.2\xi^2}} \leq 0.1$$

Thus,

$$\xi \geq 0.55$$

From Equation 13.7

$$c = 2\xi\sqrt{mk} \geq 49.2\,\text{Ns/m}$$

13.6 CONCEPTUAL QUESTIONS

1. What is meant by 'impact load' and why is it important in stress analysis?
2. Explain how an impact factor can be used to calculate deformation and stress of a structure subjected to impact loads.
3. Can an impact factor be used for structures subjected to plastic deformation and why?
4. The mid-span stress of the beam (figure below) subjected to the failing mass at B can be calculated by $K_d\sigma_{st}$, where $K_d = 1 + \sqrt{1 + 2h/\Delta_{st}}$. For the calculation of Δ_{st}, which one of the following statements is right:

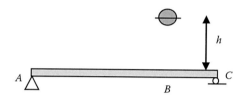

   a. Δ_{st} is the deflection of the beam at B when the mass is applied statically at B.
   b. Δ_{st} is the deflection of the beam at the mid-span when the mass is applied statically at B.
   c. Δ_{st} is the deflection of the beam at the mid-span when the mass is statically applied at the mid-span.
   d. Δ_{st} is the deflection of the beam at B when the mass is applied statically at the mid-span.

5. The two beams shown in figure below are made of the same material and have identical cross section at the fixed ends. When they are subjected to the same impact load at the free ends, which one of the following is correct for evaluating the maximum bending stresses at the fixed ends of the beams:

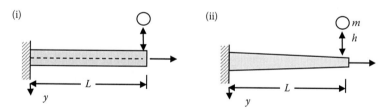

   a. $(\sigma_d)_i < (\sigma_d)_{ii}$.
   b. $(\sigma_d)_i = (\sigma_d)_{ii}$.
   c. $(\sigma_d)_i > (\sigma_d)_{ii}$.
   d. They are not comparable.

6. What are the differences between free vibration and forced vibration?
7. What is meant by 'frequency' and 'period' of vibration?
8. What is the difference between 'natural frequency' and 'natural circular frequency'?
9. What is meant by 'amplitude' of vibration?
10. What is meant by 'resonance'?
11. When an elastic system is under forced vibration condition, which of the following measures can be used to avoid resonance?
   a. Increase damping of the system.
   b. Reduce damping of the system.

c. Make sure that the frequency of the driving force is significantly different to the frequency of the elastic system.
d. Use a different material.

12. A beam is designed to support a motor as shown in the figure below. The beam can be supported in three different ways, that is, (i) simply supported at both ends; (ii) fixed at one end and pinned at the other; and (iii) fixed at both ends. Assume that the rotating part of the motor has an eccentric and the rotational speed increases continuously from zero.

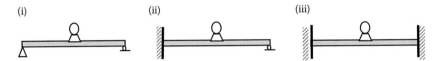

(i)   (ii)   (iii)

Which one of the statements below is correct?
a. If resonances occurs, case (i) will start first.
b. If resonances occurs, case (ii) will start first.
c. If resonances occurs, case (iii) will start first.
d. They will start resonance simultaneously.

13. How can 'DAF' be used in vibration design?

13.7 MINI TEST

PROBLEM 13.1

Consider a vertical linear spring struck by a falling mass m from height h. Show that under the impact, the contraction of the spring is

$$\left(1 + \sqrt{1 + \frac{2h}{\Delta_{st}}}\right)\Delta_{st}$$

where Δ_{st} is the contraction of the spring if the mass is placed on the top of the spring statically. (Hint: assume that the potential energy of the mass is completely converted into the strain energy of the spring.)

PROBLEM 13.2

The two columns shown in figure below are subjected to the same impact from the top ends. Which one of the following answers is correct, where K_d and σ_d are the respective impact factors and dynamic compressive stress in the columns due to the impact?

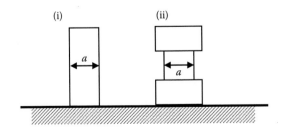

(i)   (ii)

a. $K_d^{(i)} < K_d^{(ii)}$ and $\sigma_d^{(i)} < \sigma_d^{(ii)}$

b. $K_d^{(i)} > K_d^{(ii)}$ and $\sigma_d^{(i)} > \sigma_d^{(ii)}$

c. $K_d^{(i)} < K_d^{(ii)}$ and $\sigma_d^{(i)} > \sigma_d^{(ii)}$

d. $K_d^{(i)} > K_d^{(ii)}$ and $\sigma_d^{(i)} < \sigma_d^{(ii)}$

PROBLEM 13.3

(a) Describe the free vibration responses of elastic structures that are, respectively, 'under-damped', 'over-damped' and 'critically damped'; (b) figure below shows the amplitude–frequency ratio curves of a simple vibration system. Briefly describe the features of the system.

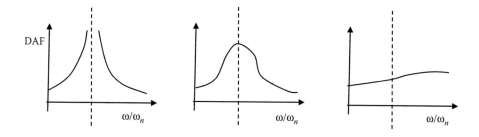

PROBLEM 13.4

A one DOF vibration system is shown in figure below, including a mass, a linear spring and a linear damper. The spring and damper are in parallel and connect the mass to the foundation in the horizontal direction. For this dynamic system, F is the external force applied on the mass directly, the stiffness coefficient k of the spring is 10,000 N/m and the mass m is 100 kg.

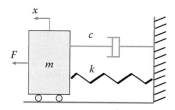

If the damping coefficient c of the damper is 600 Ns/m, calculate the natural frequency and damping ratio of the system.

a. If the required damping ratio of the system is 0.55, what is the damping coefficient c of the damper?

b. Assuming the damping coefficient c of the damper is zero and the external force F is $F = 100 \sin(5t)$ N, calculate the magnitude of the force transmitted to the foundation, and comment on the result.

c. Assuming the damping coefficient c of the damper is zero and the external force is $F = 100 \sin(5t)$ N, re-design the spring stiffness value to ensure that the magnitude of the force transmitted to the foundation is always less than 50 N.

PROBLEM 13.5

Consider the simple shear frame shown in figure below.

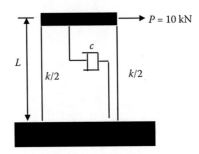

The frame is pulled to a distance of 100 mm to the right by the applied force of 10 kN and released. Following the release the time taken for the mass to return to its maximum displacement to the right is 2 s with a reduced amplitude of 80 mm:

a. Calculate f, ω_n, k, m, ξ and predict the amplitude of the frame after six cycles.
b. If a periodic horizontal force $F = P \sin \omega t$ is applied to the floor, what is the range of the angular velocity of the force such that the dynamic amplitude of the frame is less than 1.5 times of the maximum static displacement caused by P (DAF ≤ 1.5).

Index